SYSTEMS ENGINEERING SIMPLIFIED

SYSTEMS ENGINEERING SIMPLIFIED

ROBERT CLOUTIER | CLIFTON BALDWIN | MARY ALICE BONE

CRC Press
Taylor & Francis Group
Boca Raton London New York

CRC Press is an imprint of the
Taylor & Francis Group, an **informa** business

CRC Press
Taylor & Francis Group
6000 Broken Sound Parkway NW, Suite 300
Boca Raton, FL 33487-2742

Printed on acid-free paper
Version Date: 20140904

International Standard Book Number-13: 978-1-4987-0668-1 (Paperback)

Library of Congress Cataloging-in-Publication Data

Cloutier, Robert.
    Systems engineering simplified / authors, Robert Cloutier, Clifton Baldwin, Mary Alice Bone.
        pages cm
    Includes bibliographical references and index.
    ISBN 978-1-4987-0668-1 (paperback)
    1. Systems engineering. I. Baldwin, Clifton. II. Bone, Mary Alice. III. Title.

TA168.C58 2015
620.001'171--dc23   2014033163

Visit the Taylor & Francis Web site at
http://www.taylorandfrancis.com

and the CRC Press Web site at
http://www.crcpress.com

Contents

List of Illustrations

List of Tables

Foreword: The Context for Systems Engineering

Why should we bother to study systems engineering? Quite simply, because it is the single most important and pervasive discipline for the time in which we live. Today, at the dawn of the great Information Revolution, we are in the throes of a transformative leap in productivity as we experience the application of digital technology to just about everything in life. The impact of this huge productivity change (perhaps one or maybe two orders of magnitude greater than the one that characterized the Industrial Revolution) is redefining how we think about the family, the state, and our work. Today's revolution is all *about* systems: just as the exigencies of the Industrial Revolution forced the creation of mechanical, and later, electrical, chemical, and nuclear engineering disciplines, so this one demands the development of systems engineering and systems engineers.

This is, of course, merely the most recent of a series of such transformative events in human experience, each one resulting in dramatic positive changes to life span, health, and quality of life. Most of the time, productivity increases have resulted from small continuous improvements to tools and techniques: replacing fire-hardened wood points with bone tips on a spear, for example, and then progressively moving to chipped stone, pressure flaked chert, copper, bronze, iron, and steel projectile points. Occasionally, however, productivity changes are monumental and overwhelming—game changers that have millennial consequences. The most important of these earlier advances have been (in my view, at least) cooking, farming, writing, and steam power.

The control of fire was important primarily because it led to the invention of cooking, dramatically increasing the calories available to hunter-gatherers from both plant and animal biomass. It is believed that this resulted in dramatic increases in brain size, including the development of areas of the brain related to speech processing. With increased cognition came knowledge and insight about seasonal crop rotation and animal migrations. Experimentation with plant and animal management was subsequently responsible for the initiation of farming, with another huge productivity payoff. Population growth made possible by farming in turn led humans to congregate into cities. City life and its barter system of exchanges among specialist artisans led almost inevitably to the development of writing, and the consequent freedom from the difficult and uncertain process of knowledge transmission through word of mouth. This agriculture and artisanship-based model of human life lasted virtually unchanged for around 7 millennia, until finally in the middle years of the 19th century, the invention of steam power irrevocably changed the landscape yet again.

The productivity revolution of the mid-1900s was astonishing and unprecedented in recorded history (not surprising since the last of these revolutions had been the one that gave us recorded history to begin with!). While the Industrial Revolution played itself out over the 100 years running from roughly 1840 to 1950, the pace of change was unnerving, as the application of steam power (and later, other forms of energy, such as oil, gas, and hydroelectric) to just about every aspect of life upended expectations about everything from the nature of the family, to work, to the role of public society.

The unique impact of today's Information Revolution is that for the first time in history—or at least to a greater extent than ever before in history—it is necessary for ordinary citizens, not just expert practitioners, to have a sense of the theory behind a branch of engineering, in this case, systems engineering. Without it, we risk becoming baffled by the behavior of engineered *things* that have a deep impact on our lives: Why do our devices behave the way they do? Why do so many seemingly straightforward problems defy solution in the face of Herculean efforts by earnest and well-meaning people? Why do so many outcomes in our lives feel counterintuitive? It is because they all involve systems, and these in turn are behaving in accordance with the laws of systems.

An understanding of the principles of systems engineering is the most powerful skill an average citizen can have in today's world. It provides an inoculation against conspiracy theories, Internet hoaxes, and telemarketing scams. And it most definitely helps us to remain calm in the face of baffling behavior on the part of the institutions and *systems* that we are all embedded in. Not that we are any less inconvenienced, but at least we can understand that whatever is bothering us is not the fault of the hapless humans with whom we are trying to deal, but a result of how the system is put together.

The first steps on this journey must include an understanding of the underlying nature of the system as an object of study, knowledge of the vocabulary of systems engineering, recognition of the various tools available to systems engineers for the study of systems, and a working grasp of the process we use to design, build, test, operate, and maintain systems in the everyday world. These are precisely the topics covered in this book, and they are purposely covered in a way that does not require sophisticated technical training or complicated mathematics. The authors have provided us with an essential fundamental review of the principles of systems and systems engineering designed to illuminate the subject for a general audience. This book serves as a primer that will help you understand the broad outlines of how systems work, and what systems engineers do to design, build, and operate them. It is an indispensable guide to the perplexing realities of the 21st century.

Dr. Wilson N. Felder
Castle Point on Hudson

Dr. Felder is director of Washington Operations and a distinguished service professor at the Stevens Institute of Technology. He is the former director of the William J. Hughes Technical Center in Atlantic City, New Jersey, the FAA's national research and test laboratory.

Preface

Systems engineering is about thinking—thinking about things in a systematic manner. It first addresses the question of "Why are we doing this?" It then moves to understanding how the idea will be used in the current environment. Systems engineers also ask questions like "Did we build the right thing? Does it do what the end user wants?" Without these answers, one runs the risk of producing the next product that the market does not want. Systems engineering is sometimes characterized as a new discipline since it often goes unidentified. But, in fact, many companies are practicing systems engineering. They just do these things under a different name.

In the first chapter we demonstrate that engineers in commercial and government programs began to recognize that the world was becoming more complex, and therefore a more holistic approach was needed when thinking about these newer systems. Subsequent chapters break down the systems engineering life cycle, describing in the simplest of terms what should be done along the development process.

This book is not meant to be a textbook, but instead, a gentle introduction to systems engineering. The intended audience for this book was the undergraduate engineering students in the National Science Foundation (NSF) Center for Compact and Efficient Fluid Power. These undergraduate students are not systems engineers. However, they need a better understanding of systems engineering while they consider their applications of fluid power into broader systems. However, as others have reviewed this manuscript, they have suggested this material would be very helpful to a much broader audience.

When I was requested to write this manuscript, I reached out to a couple of my colleagues that also have considerable systems engineering experience—Dr. Clifton Baldwin and Mary Bone. So, this book represents a composite of systems engineering, as practiced on large government programs and commercial companies.

Although the views and material in this book are our own, we did get some suggestions and help in reviewing early drafts of the book. We thank Jody DeMarco, PE, and Mike Paglione for their comments and suggestions from non-systems engineering points of view. In addition, we thank systems engineers Ji'on Crump, Tony Long, Stacy Cornish, Lakaisha Ajaegbulemh, Marie Kee, and Dr. George Gardner for their input as they strive to improve systems engineering within the Federal Aviation Administration (FAA). Last but not least, we thank Susan A.M. Baldwin for her artistic talents when we needed a custom-made graphic.

We hope you enjoy this simple introduction to systems engineering.

Rob Cloutier, Clif Baldwin, and Mary Bone

About the Authors

Robert Cloutier is an associate professor of systems engineering for the School of System and Enterprises, Stevens Institute of Technology. He serves on the Scientific Advisory Board for the National Science Foundation Engineering Research Center for Compact and Efficient Fluid Power. Rob is the current president of the Delaware Valley Chapter of International Council on Systems Engineering (INCOSE). His research interests are focused on improving concept engineering and the way concepts of operations (CONOPS) are created, applying patterns while architecting complex systems, and model-based systems engineering. Rob has over 20 years of industry experience prior to joining academia. Industry roles included principal systems engineer, system architect, lead systems engineer, and engineering project manager for major defense contractors.

Rob graduated with a BS from the U.S. Naval Academy, an MBA from Eastern University, and a PhD in systems engineering from Stevens Institute of Technology.

Clifton Baldwin is a senior systems engineer with the Federal Aviation Administration and a postdoctoral researcher at the Stevens Institute of Technology. Clif is the southern New Jersey regional director for the Delaware Valley Chapter of INCOSE. He has over 20 years of experience working in software and systems engineering. His research interests include system of systems and complex systems modeling. He holds a BA degree in mathematics from Rutgers University, an MS degree in information systems from Johns Hopkins University, and a PhD in systems engineering from Stevens Institute of Technology. In addition, Clif is certified as a Project Management Professional (PMP) by the Project Management Institute (PMI) and as an Expert Systems Engineering Professional (ESEP) by INCOSE.

Mary Bone is a doctoral candidate at Stevens Institute of Technology. She holds a BS in aerospace engineering from Missouri University of Science and Technology and an MEng in systems engineering from Iowa State University. She has worked as a systems engineer for GE, BAE, Rockwell Collins, and Dell. She has held roles in system design, requirements, verification, validation, and systems engineering steering teams. She has also been a research assistant at Stevens Institute of Technology performing research on system architectures, modeling, and evaluation. Her current research is in system architecture complexity and entropy.

1 Introduction

1.1 OVERVIEW

When Bob bought his new 2000 Cadillac Sedan DeVille DTS (Figure 1.1), one of the first things he wanted to do was upgrade the factory sound system. He took it to Kenney's Allston shop, Sound in Motion, and selected a new head (radio) and asked Derek Kenney to install it in the new car. When Kenney completed the removal of the factory radio and installed the new head, the new system sounded great. However, he soon discovered that pulling out the original radio resulted in the car's air conditioning, alarm, and computer diagnostics systems to stop working. The only way Kenney could get the new head to work without disabling the A/C, alarm, and computer diagnostics was to install the old radio in the trunk. "The guy actually has two radios in his car, one he listens to and one to keep the car working," Kenney said.*

What happened? When the A/C engineers were designing their system for this vehicle, they needed a little more processing power and noticed the radio had excess capacity in its electronic control unit (ECU). Rather than adding another ECU for the A/C, they used the excess capacity of the radio, thus tightly coupling the radio with the A/C. The same approach was used when the computer diagnostics engineers needed a bit more capacity for the alarm system. The result? Four unique subsystems, that have no relationship to one another, became tightly coupled. Whether that was to save a few cents in production costs or not, it represented a lack of a systems view—someone not watching each of the parts in the context of the whole. In the end, the old radio ended up in the trunk just to keep the car running properly.

Systems engineering is an engineering approach that provides an understanding of the interaction of individual parts that operate in concert with one another to accomplish a task or purpose (Figure 1.2). Today the title of *systems engineer* is used within a number of different professions, ranging from computer systems administrators to engineers who design spacecraft. While these professions have many differences, they share a goal of looking at a system as a whole.

The impact of these different applications of the term *system* is that it makes it difficult to arrive at a single accepted definition for systems engineering, but there are some commonalities among the descriptions. A very minimalistic approach to systems engineering is "good engineering with special areas of emphasis" (Blanchard and Fabrycky 1998, 23). Here "good" means viewing a system as a whole from a top-down approach and addressing all phases of the system life cycle. Those areas would include a strong effort on the initial definition of system requirements, system validation, and taking an interdisciplinary approach to ensure all aspects of design are addressed.

---

* Adapted from a story by Keith Reed, *Boston Globe* staff, August 23, 2004, http://www.boston. com/business/technology/articles/2004/08/23/to_installers_of_car_stereos_auto_systems_sound_ fishy?mode=PF.

FIGURE 1.1 A 2000 Cadillac Sedan DeVille DTS. (From http://en.wikipedia.org/wiki/Cadillac_de_Ville_series#mediaviewer/File:Cadillac_Deville_--_10-30-2009.jpg.)

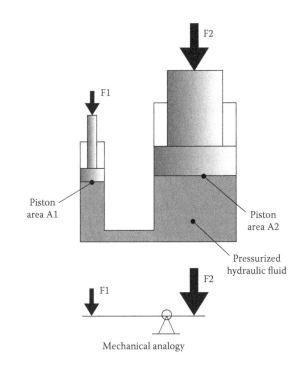

FIGURE 1.2 A simple hydraulic system. (From http://en.wikipedia.org/wiki/Hydraulic_drive_system#mediaviewer/File:Hydraulic_Force_Torque_275px.png.)

The International Council on Systems Engineering (INCOSE) states that systems engineering is "an interdisciplinary approach and means to enable the realization of successful systems. It focuses on defining customer needs and required functionality early in the development cycle, documenting requirements, and then proceeding with design synthesis and system validation while considering the complete problem: operations, cost and schedule, performance, training and support, test, manufacturing, and disposal. Systems engineering considers both the business and the technical needs of all customers with the goal of providing a quality product that meets the user needs" (INCOSE 2010).

Systems engineers come from many different engineering or technical backgrounds. Some started as electrical engineers, some as mechanical engineers, or any number of backgrounds. Some of the best systems engineers we have met have an educational background in astrophysics. These individuals are excellent systems engineers because of their ability to imagine how planets and stellar objects work in concert with one another. Systems engineers are problem solvers who are skilled in identifying the true problem or challenge before they ever begin trying to solve it. Once the root problem or challenge has been identified and understood, systems engineers use proven processes, analysis techniques, and management technology to describe the problem, develop an architecture and design, build the solution, then make sure they built the right system and that it solves the original problem. This process is known as the system development life cycle. There are also a number of different descriptions of the system life cycle, but they all begin with the conception of a new system, or a reason to engineer or modify an existing system. The life cycle proceeds through the gathering of requirements, architecture and design, production, utilization and support, and finally retirement and disposal. Throughout the life cycle, systems engineering considers the whole system as well as its elements.

1.2   DISCUSSION OF COMMON TERMINOLOGY

As in any other field, words and definitions are important. This section is meant to provide a straightforward discussion on some of the terms that will be used throughout this handbook.

> **System:** Sometimes we ask, "Who is buried in Grant's tomb?" Of course it is Grant. Well, systems engineering is the understanding of systems. So what is a system? This question is harder to answer than it first appears. Why? When thinking about systems, virtually anything might be considered a system, depending on the inquisitor's perspective and background. Consequently, there are varying definitions. In order to provide a well-rounded view, the following definitions are provided. These descriptions are by no means a definitive set, as the number of authoritative attempts is vast.
>
> > **Definition 1:** A system is an assemblage or combination of elements or parts forming a complex or unitary whole, such as a river system or a transportation system; any assemblage or set of correlated members, such as a system of currency; an ordered and comprehensive assemblage of facts, principles, or doctrines in a particular field of knowledge or thought, such as a system of philosophy; a coordinated body of methods or a complex scheme or plan of procedure, such as a system of organization and management; any regular or special method of plan

of procedure, such as a system of marking, numbering, or measuring (Blanchard and Fabrycky 1998).

Definition 2: An interacting combination of elements viewed in relation to function (INCOSE 2010).

Definition 3: An integrated composite of people, products, and processes that provide a capability to satisfy a stated need or objective (Mil-Std 499B).

Definition 4: A set of interrelated components that interact with one another in an organized fashion toward a common purpose (NASA 2007).

Definition 5: A construct or collection of different elements that together produce results not obtainable by the elements alone (INCOSE 2006).

If we were to create an aggregation of these definitions, we might think of a system as:

A system is a set of interrelated parts that work together to accomplish a common purpose or mission.

Based on these definitions, *we are also going to adopt the principle that a system must have a mission that it can complete without the aid of another system.*

This assumption has nothing to do with the complexity of the system, but rather, can the system perform the intended mission independently? Although the concept of a system touches many fields, we will focus on systems that are engineered. For instance, using this definition, a bicycle is considered a system (Figure 1.3)—it is a set of interrelated parts that work to transport a rider to a destination. In this case, the mission is to transport the rider. As you will find later in this book, the rider is both part of the system and a stakeholder of the system.

System of interest (SoI): In order to easily apply the systems engineering principles outlined in this monograph at all levels throughout a system (e.g., subsystem level, module level, component level), the systems engineering community has introduced the notion of system of interest. This concept allows a systems engineer to think of any part of any system as a system of its own. If one treats that part as a black box, and considers all the interfaces as external interfaces, you can then apply these systems engineering principles to that black box.

System of systems (SoS): Now that we have considered a system, let's think about what happens when there are a number of systems working together. Examples might be the individual airplanes moving people across the skies between destinations. This is often called a system of systems (SoS). There are two different mainstream viewpoints on SoS (Baldwin et al. 2011). On one hand, a SoS may be simply a matter of perspective where every system can be considered part of a larger system (Ackoff 1981). In this case, there is nothing exceptional about a SoS other than its scope. On the other hand,

FIGURE 1.3 The bicycle is a system. (Courtesy of PRESENTERMEDIA.)

a SoS may be a distinct entity with a unique set of characteristics and traits. Although both views have their merits, it is prudent for the systems engineer to understand the unique characteristics and traits. For the sake of this discussion, we will consider a SoS as fundamentally a system, as its name suggests, that exists as a composite system whose diverse constituent systems are independent, dynamically connected, and contribute to a unique overall goal of the system as a whole. More time will be spent on this type of system and the challenges it imposes in Chapter 8.

System life cycle: The term *life cycle* refers to the entire spectrum of activities for a given system, starting with identification of a need and extending through design and development, production and construction, operational use, sustainment of support and system retirement, and eventually, disposal. Different systems may have different life cycles according to the nature, purpose, use, and prevailing circumstances of the system. In every case, each stage of a life cycle has a distinct purpose and contribution to the system as a whole (ISO-15288 2008). The systems engineer must consider every stage when planning and executing the various activities throughout the system's life cycle.

Requirement: A requirement is any expectation, desire, or need based on a true or perceived deficiency. The system must satisfy this expectation, desire, or need to be considered a good system. Requirements might address what the system does (transport a passenger from point a to point b), basic operating characteristics (attain a maximum speed of 100 miles per hour), safety (provide a passive restraint for each passenger), and reliability (require regular oil changes at a frequency exceeding 3000 miles), as well as many others. In Chapter 5, we will discuss at length stakeholder requirements and system requirements. While these terms have their own meanings, the underlying intention remains: they are necessary or essential conditions for a system to accomplish its mission.

Stakeholder: A stakeholder is any individual, another system, or even an organization (such as a government agency) with a legitimate interest in the system (INCOSE 2010). Normally we think of the stakeholders as the people who will be using the system, such as the user of a kiosk (Figure 1.4).

FIGURE 1.4 A stakeholder of a kiosk-based system. (Courtesy of PRESENTERMEDIA.)

However, stakeholders can also be an organization or individual who has an interest in the outcome of the engineering of a system (EIA-632 1999). Although project personnel often use the term *stakeholder* to refer to those not directly involved with the engineering of a system, it does include the project manager, systems engineers, and other developers, as well as the end users, system operators, sales personnel, and management. If the system requires certification or governance by a government agency, that agency is also a stakeholder. Of course, there are many instances where all stakeholders are not included in the engineering of a system, but involvement in the engineering of a system is different than interest in it. To demonstrate this point, you may not have been asked about the design of a new cell phone, but you are a stakeholder once you own one of the phones.

Subsystem: A subsystem is the name given to a part of the system that is a significant part of the system hierarchy when there are two or more levels in the system hierarchy (Blanchard and Fabrycky 1998). Within a rail transportation system, the trains and signaling system would be two examples of subsystems, although in different contexts these subsystems may be considered systems. Continuing our bicycle example, the derailleur on a multi-speed bike may be considered a subsystem (Figure 1.5).

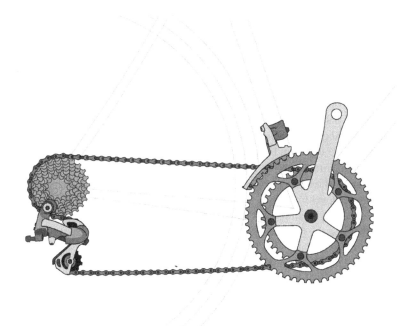

FIGURE 1.5 A derailleur on a bicycle may be a subsystem. (From http://en.wikipedia. org/wiki/Derailleur_gears#mediaviewer/File:Campagnolo_Super_Record_rear_derailleur_1983.jpg.)

Component: A component is a subset of a system that has been allocated certain functions and requirements of the system (Buede 2000). One of the individual gears seen in Figure 1.5 would be considered a component of the bicycle system.

1.3   THE CASE FOR SYSTEMS ENGINEERING

As we established earlier, systems engineering is focused on viewing a system as a whole—a holistic view. Systems engineers are trained to take a methodical approach to understanding a system—from cradle to grave. This approach includes considering how the system will be built, how the system will be supported once it is in the customer's hands, and how it will be disposed of when the customer is done with it.

An example that demonstrates the importance of systems engineering is the Sprint spacecraft, which was designed to make very short flights in space. When the spaceship was first envisioned, no one expected that it would be necessary to change the spacecraft batteries in orbit. Therefore, there was no need to consider rechargeable batteries, and the designers created a support requirement for the purchase of additional batteries to support testing. Based on the requirements, rechargeable batteries seemed to be an extravagance. Instead, lithium ion (Li) batteries were used, as they met the requirement for the battery life. However, due to the volatility of Li batteries, the battery compartment was designed with over 100 screws to secure the battery pack.

A problem arose that extended the planned testing phase. Because of the longer test phase, the Li batteries needed to be replaced countless times. Each time they had to be replaced, all 100 screws had to be removed and then reinstalled, thus adding yet more time to the test phase. The Li battery packs were very expensive, build-to-order items, and the project quickly burned through the entire life cycle inventory of batteries during the test phase.

The Defense Acquisition University studied this problem in the early 1990s and produced the graph shown in Figure 1.6. It shows a number of items of concern to the systems engineer. First, it shows that roughly 70% of the cost to develop a system is committed during the concept phase. What this means is that engineering decisions made during the development of the system concept, the system requirements, and the system architecting steps will drive 70% of the final cost to produce the entire system. By the time the final design is complete, but before development begins, 85% of the costs will be committed.

Second, this graph shows that mistakes made during the concept phase (and not fixed during that phase) will cost three to six times more to fix in the design phase. If a mistake is not found until production and test, it will cost 500 to 1000 times more money to fix than if it were found in the phase it was generated. The longer a problem goes uncorrected, the worse the problem will become. Therefore, reducing the risk associated with new systems or modifications to complex systems continues to be a primary goal of the systems engineer.

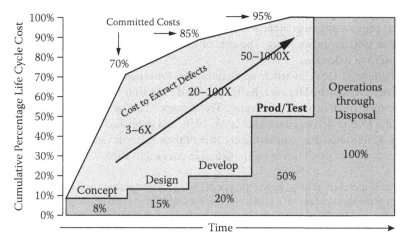

Committed Life Cycle Cost against Time
(From Defense Acquisition University 1993)

FIGURE 1.6   Seventy percent of the product costs are allocated in the first 10% of the project life cycle. (From Defense Acquisition University, 1993.)

1.4   A BRIEF HISTORY OF SYSTEMS ENGINEERING

While some will say that systems engineering is a new engineering discipline, that is actually incorrect. Although systems engineering is relatively newer than classical engineering, such as mechanical and civil, it has existed as a discipline since before World War II. For example, Arthur Hall wrote an early textbook* published in 1962 detailing the practice of systems engineering at that time. He noted in the preface that "the growing recognition of the need for systems engineering over the past decade has been attended by the need for philosophical foundations." He stated that his book was intended "to increase awareness and understanding of systems engineering as a process, and to sharpen definitions and approaches to the main recurring problems of the process-problem definition, goal setting, systems synthesis, systems analysis and choice among alternative systems."

Hall published a composite view of the systems engineering process, as it was defined in 1962. Prior to Hall's book, an AIEE fellow, E.W. Engstrom, published a paper in the journal *Electrical Engineering*.† It was titled "Systems Engineering— A Growing Concept." Engstrom was an engineer at the Radio Corporation of America (RCA). In his 1957 paper, Engstrom stated, "The task of adapting our increasingly complex devices and techniques to the requirements and limitations of the people who must use them has presented modern engineering with its greatest challenge.

---

* A.D. Hall, *A Methodology for Systems Engineering*, Princeton, NJ: Van Nostrand, 1962, p. 139.
† E.W. Engstrom, Systems Engineering—A Growing Concept, *Elec. Eng.* 76(2), 113–116, 1957.

To meet this challenge, we have come to rely increasingly during recent years upon the comprehensive and logical concept known as systems engineering."

According to Hall, AT&T Bell Labs began a 3-year graduate degree program in systems engineering in conjunction with MIT in 1948. In 1950, MIT offered its first course in systems engineering. The goal of this program was to provide consistency in transmission designs.

In November 1950, an article appeared in the *Proceedings of the Royal Society B*. The article, written by Mervin J. Kelly, was titled "The Bell Telephone Laboratories— An Example of an Institute of Creative Technology." This paper was presented in the form of a lecture delivered on March 23, 1950, and published later that year. In this article, Kelly stated that approximately 10% of their scientific and technical staff was allocated to systems engineering (p. 426). He goes on to state:

> Its staff members must supply a proper blending of competence and background in each of the three areas that it contacts: research and fundamental development, specific systems and facilities development, and operations. It is, therefore, largely made up of men drawn from these areas who have exhibited unusual talents in analysis and the objectivity so essential to their appraisal responsibility.

The last example we will present is the Jet Propulsion Laboratory (JPL) in California. Engineers at JPL were responsible for the development of the Army Corporal sounding rocket in 1945. The lack of engineering process consistency resulted in a lack of reliability in their systems. Therefore, they began a process to transition their approach from a research mindset to an engineering methodology. Their first methods were based on "tribal knowledge," or rather, a standardization of their then currently accepted domain knowledge.

So, we return to the claim introduced in the beginning of this section that systems engineering is not a new discipline. It has been in practice at least since 1945, which represents over 60 years of experience and use. Systems engineering is hardly a young discipline.

1.5   SYSTEM EXAMPLES

What makes up a system? What are the rules that enable something to be referenced as a system? Consider the items in Figure 1.7. Any of these might be considered a system. While some are more complex than others, they can all be considered a system, by definition. They have interrelated parts working together toward their own goal or mission. Some may argue that a desktop stapler is not a system because it is not complex enough. However, what is complex to one person may not be complex to someone else.

As we discussed earlier in this chapter, there are as many definitions as there are opinions to answer the question of what a system is. And, we stated that a system must have a mission that it can complete without the aid of another system. So, let us look at an example. The Boeing 747 is a system (Figure 1.8). It has many complex parts, including the jet engine (Figure 1.9). Some would say the engine is a system, too. However, if one were to remove that engine and place it on the

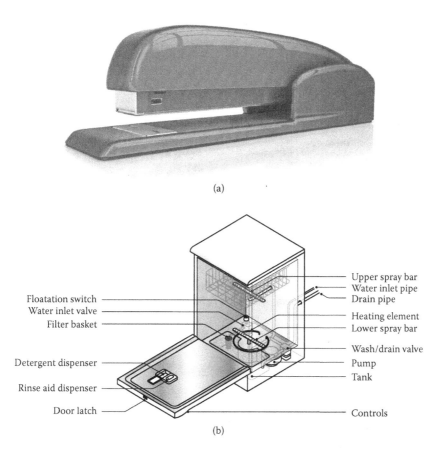

(a)

Floatation switch
Water inlet valve
Filter basket

Detergent dispenser

Rinse aid dispenser

Door latch

Upper spray bar
Water inlet pipe
Drain pipe

Heating element
Lower spray bar

Wash/drain valve
Pump
Tank

Controls

(b)

(c)

FIGURE 1.7   Very different systems. (a) Clip art; (b) from www.shutterstock.com 161660264; (c) from http://en.wikipedia.org/wiki/Excavator#mediaviewer/File:Kettenbagger_CAT_325C_LN.jpeg (creative commons (cc) license).

FIGURE 1.8 The Boeing 747, first operated by Pan Am. (From http://en.wikipedia.org/wiki/Boeing_747#mediaviewer/File:Pan_Am_Boeing_747_at_Zurich_Airport_in_May_1985.jpg.)

FIGURE 1.9 Pratt and Whitney engine used on the Boeing 747. (From http://en.wikipedia.org/wiki/Pratt_%26_Whitney_PW4000#mediaviewer/File:Aircraft_part_-_engine_01b.JPG.)

runway, could that engine perform the mission for which it was designed, that is, provide forward thrust to a Boeing 747? No. The fuel for the engine is in the airplane wing. The power to start that jet engine is on the aircraft. Yet, the aircraft can take off, fly somewhere else, and land without the aid of another system. Yes, that means the pilot is part of the system (Figure 1.10). Now, there are other rules

FIGURE 1.10   The pilot is part of the system. (Courtesy of PRESENTERMEDIA.)

and regulations that state that pilots need to file a flight plan with the FAA, need clearance from the tower, etc. However, the pilot could choose to ignore those requirements and fly the aircraft. The jet engine on the tarmac cannot propel the aircraft until it is attached to the wing. Therefore, the jet engine is a subsystem of the aircraft—albeit a very complex and sophisticated subsystem. In Chapter 4 we address why systems engineering is still critical to all aspects of the system and its constituent parts.

1.6   SUMMARY

This chapter has presented a number of definitions and concepts. These ideas may seem confusing, but consider the following quote from Russell Ackoff (1981). He is a leader in systems thinking and describes a system as

> a whole that cannot be divided into independent parts without losing its essential characteristics as a whole. It follows from this definition that, a system's essential defining properties are the product of the interactions of its parts, not the actions of the parts considered separately. Therefore, when a system is taken apart, or its parts are considered independently of each other, the system loses its essential properties. Furthermore, when performance of each part taken separately is improved, the performance of the system as a whole may not be, and usually isn't.

Mind mapping is a diagrammatic approach to thinking about a concept. Figure 1.11 represents one perspective on the practice of systems engineering developed by a team of systems engineers during a lunch-and-learn. That concept is placed in the center, and then limbs are drawn off that "trunk" to describe the major limbs, or thoughts about the main concept.

Applying the same mind mapping approach, Figure 1.12 represents a mind map of the concepts relating to a system. While we have not discussed each of these concepts, they are important to the practicing systems engineer. What is clear from this chapter is that the definition of a system, and therefore the practice of systems

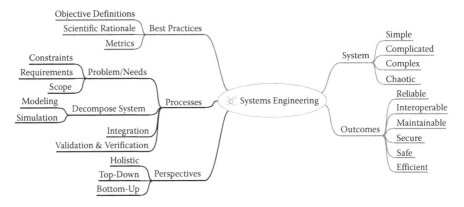

FIGURE 1.11   Mind map representing one perspective of the practice of systems engineering.

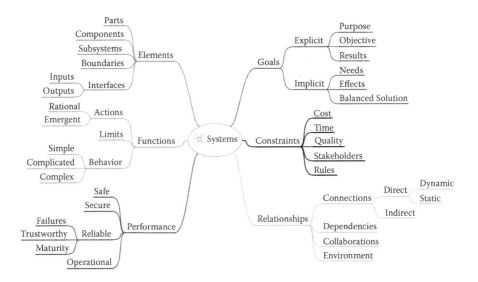

FIGURE 1.12   Mind map of the concept of systems.

A system is a set of interrelated components which interact with one another in an organized fashion toward a common purpose. The components of a system may be quite diverse, consisting of persons, organizations, procedures, software, equipments, or facilities. The purpose of a system may be as humble as distributing electrical power within a spacecraft or as grand as exploring the surface of Mars. NASA Systems Engineering Handbook, SP-610S, June 1995	*A "system" is a construct or collection of different elements that together produce results not obtainable by the elements alone. The elements, or parts, can include people, hardware, software, facilities, policies, and documents; that is, all things required to produce system-level results. The results include system-level qualities, properties, characteristics, functions, behavior, and performance. The value added by the system as a whole, beyond that contributed independently by the parts, is primarily created by the relationship among the parts; that is, how they are interconnected.* NASA Systems Engineering Handbook, SP-2007-6015, Revl, December 2007 (*Quoting Rechtin, Systems Architecting of Organizations: Why Eagles Can't Swim*)

FIGURE 1.13 Evolving definitions of a system as defined by NASA, separated by 12 years. (Courtesy of NASA.)

engineering, continues to evolve. The two quotes shown in Figure 1.13 show that the definition and understanding of systems engineering continue to evolve as our systems become more complex. These quotes are from two different versions of the *NASA Systems Engineering Handbooks* published over a 12-year period (1995–2007).

2 The System Life Cycle

Here systems engineering is focused on addressing *why* a system is needed, *what* the system must do, and then *how* the system will accomplish the tasks over the entire system life cycle from conception to disposal (Figure 2.1). Therefore, understanding the system life cycle is important for systems engineers since many times only the design phase is addressed in introductory engineering material. Being able to implement a solution across the system life cycle is what forces systems engineers to understand other areas of study, such as management, manufacturing, marketing, sales, and customer support, since these areas all intersect with the system in the life cycle. Figure 2.2 shows the system life cycle that most systems will go through. While some systems may go through a phase of the cycle at a different rate than another system, the phases remain the same. It is also important to point out that while Figure 2.2 shows clear breaks between the phases, it almost never happens this way in real life. In reality a systems engineer must be able to break a system down into subsystems, components, or some other elements and move each piece of the system through the process while also moving the whole system through the life cycle. Taking this dual view of both the pieces and the whole is arguably one of the most important skills of a systems engineer.

So how does this life cycle look in real life? We will use a dishwasher to show the flow through a top-level life cycle (Figure 2.3). You may ask why it is important for the systems engineer to consider the entire life cycle. While many engineers are only trained in their specific area of expertise (e.g., electrical, mechanical, manufacturing, etc.), there needs to be someone to consider the entire system despite its stage of development. For example, an electrical engineer is concerned with the power system of a dishwasher. The machine has to work with household current and not electrocute anyone. The systems engineer has to ensure the machine will meet the customer needs, which may include being environmentally friendly. Think about Figure 2.3 and consider the different hats the systems engineer must wear when having discussions with the other members of the product team. An example of questions the systems engineer needs to consider and address with specific domain engineers follows:

1. Environmental engineer: Is the dishwasher environmentally friendly?
2. Human factors engineer: Is the dishwasher easy to use?
3. Software engineer: Does the software control the dishwasher properly?
4. Fluid engineer: Does the dishwasher use the proper hose connectors?
5. Electrical engineer: Does the dishwasher support both the 60-Hz standard used in the United States and the 50-Hz standard used in the rest of the world?
6. Safety engineer: Does the dishwasher sanitize the dishes to a safe bacterial level without violating the other needs?

FIGURE 2.1   Cyclical. (Courtesy of PRESENTERMEDIA.)

| Conception | Design | Production | Utilization, Maintenance, & Support | Disposal & Retirement |

FIGURE 2.2   System life cycle.

| Need more energy & water efficient dishwasher | Company sets off to create best design | Manufacturing produces new dishwasher design | Customer buys and utilizes new dishwasher | Customer replaces dishwasher and disposes through a recycling program |

FIGURE 2.3   Dishwasher life cycle.

You get the idea. While the development team should include many types of folks trained in specific areas, there must be a leader responsible for life cycle-driven decisions and pressing forward. This person is usually a systems engineer. Therefore, it is important for the systems engineer to understand the whole life cycle and not just the design part (Figure 2.4). Just because the engineer is classically taught and trained to design the very best product, it is not always the best engineered product that sells. Balancing all aspects of the system is a trait of a successfully engineered system.

The rest of this chapter will look at each of the life cycle phases in detail. We will use the Vee chart to make the discussion more relevant.

FIGURE 2.4   The systems engineer must be aware of other engineering discipline issues. (Courtesy of PRESENTERMEDIA.)

2.1   MANAGING SYSTEM DEVELOPMENT—THE VEE MODEL

While there are different systems engineering methods, they appear to be different flavors of the same process. They can be "generalized as State the problem, Investigate alternatives, Model the system, Integrate, Launch the system, Assess performance, and Re-evaluate" (Bahill and Gissing 1998). Stating the problem addresses the why, such as why a system is needed. Investigating alternatives and early modeling of the system address the what, as in what the system must do. The process continues with how the system will do what it must do, and then loops back to ensure the what and why were properly addressed.

The Vee model (Figure 2.5) represents the basic process for developing a system. On the left side of the Vee is the design and build phases of product development. The right side of the Vee represents the integration, verification, and validation of the parts into a complete system. The black dotted line in the middle of the Vee model is approximately where the line of responsibility is divided between systems engineering and those tasks handled by other engineering disciplines (i.e., electrical, mechanical, software, etc.). For any Vee, the phases on the right side match up with the appropriate level of requirements on the left side to ensure they are verified and validated.

The Vee model can be used on projects of differing complexity and can be iterated by moving back up or down the Vee as needed. Actually, separate Vee models

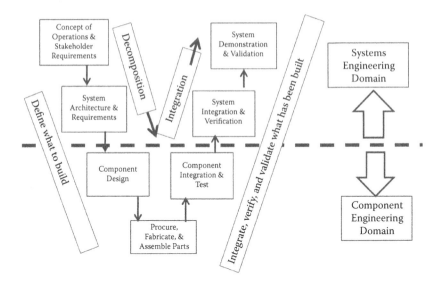

FIGURE 2.5   The systems engineering Vee model.

can be applied to each phase of the overall Vee model. For example, in the concept of operations and stakeholder requirements phase, there may be an entire Vee model to design, implement, and validate the concept of operations and stakeholder requirements.

In the next few sections the roles and responsibilities at each step of the Vee model will be discussed along with the description of each step.

2.1.1   Concepts and Stakeholder Requirements

Systems engineering commences when a need is identified in the form of a new or improved capability, or to provide a missing capability. First, the systems engineer identifies all the stakeholders associated with the need and capability. Then the systems engineer can work with the appropriate stakeholders to define the true need and not unjustified wants. This need will form the basis for an initial concept. The conception phase of the system life cycle (Figure 2.6) is the time from the initial concepts or identification of a need to the time formal design begins.

This phase includes concept definition, concept selection, proof of concept, and documenting what the stakeholder wants (customer/stakeholder requirements). The main goals during this phase should be to understand what the stakeholders really want, and the opportunities and risks of moving forward to the design phase.

The system conception phase can be the hardest phase for many because it may require the systems engineer to be a visionary and make predictions for months or years down the path. Systems engineers can find it easy to get stuck in this phase and not want to move on until the risk of moving forward is viewed to be very minor. It becomes important during the conception phase to understand what you do not know and to make sure what you do know holds true for the system. A quote

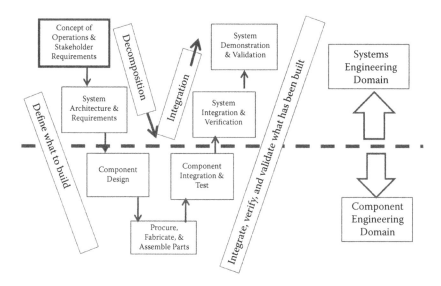

FIGURE 2.6   The first phase of the Vee.

by Mark Twain to remember during the conception phase is "It ain't what you don't know that gets you into trouble. It's what you know for sure that just ain't so." Many times early in the conception phase engineers will believe that a risk has been eliminated because they believe they have figured out the issue in every detail. Yet when design begins, the system behaves very differently as soon as an unexpected variable is changed or put into play. The systems engineer must do a good job of understanding what is known and what is not known and be careful of making assumptions that just are not true.

Starting at this early phase, the systems engineer is responsible for knowing the technical opportunities and risks of moving forward. He or she must always be on the lookout for new risks—things that would be bad if they were to occur during the development of the product. The systems engineer documents these risks and opportunities so they can be considered, and hopefully fixed or exploited, as soon as possible. One thing to remember, at every phase, is that the systems engineer is expected to interface with the management team and ensure the proposed system continues to address the real problem or need (Figure 2.7).

Upon completion of identifying and documenting the need or opportunity, the systems engineer develops an initial system concept. The financial stakeholders may require a proof of concept (maybe a prototype or model) before authorizing the project to begin in earnest.

Once the proof of concept is complete, or at least understood, work on a concept of operations (CONOPS) should begin. The CONOPS is usually a set of scenarios or use cases that demonstrate how the completed system will be used. It may be as simple as a single-page description, a storyboard, or an elaborate simulation. The CONOPS should include scenarios from all phases of the life cycle and answer the questions "How will the system be used?" and "Who will

FIGURE 2.7   Man with telescope. (Courtesy of PRESENTERMEDIA.)

use the system when?" However, the CONOPS does not describe how to develop the system or how it works. Once the systems engineer understands how the customer wants to use the system, then he or she can start writing the stakeholder requirements.

Continuing with the dishwasher example, the systems engineer identifies the operator, the service person, the house where the system will reside, the financers and managers of this project, and the developers of this system, as well as any regulatory agencies that must approve the system for household use. The operator wants to clean dishes, service the dishwasher, and install/uninstall the dishwasher. Figure 2.8 shows how this can be drawn using the Systems Modeling Language (SysML). This particular concept of operations (shown using a use case diagram) illustrates the actors of the system and demonstrates the concept of maintaining the dishwasher once it has been installed in a customer's home. The actors in this case are the operator (homeowner), service person, and the house itself. The house is an actor because it is also a system that provides something to the dishwasher system, namely, water, power, drainage, and mechanical support. The operator and house actors are involved with the actions of cleaning dishes, which include loading and unloading dishes. The service person interacts with the operator and plays a role with the actions of service, and installing and uninstalling the dishwasher. While this is a simplistic concept of operation, it is important to start identifying these types of interactions and what the system must do. The systems engineer uses these higher-level details to identify and describe the necessary lower-level interactions to continue developing the dishwasher.

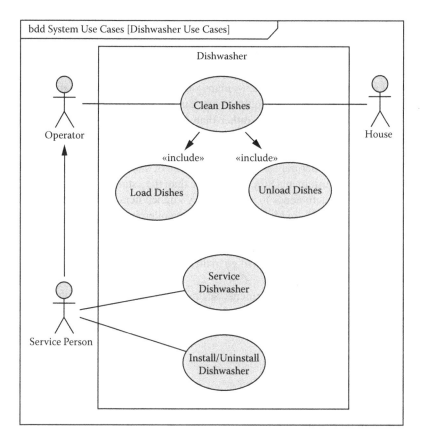

FIGURE 2.8 Dishwasher operational concept (use case).

The requirements are elicited usually through the stakeholders (management, marketing, end users, maintainers, homebuilders, etc.). A very brief sampling of those stakeholder requirements from the example can be seen below. The stakeholder requirements express what the actors and stakeholders deem the system must accomplish to be successful (functions) and how well each function should be accomplished (performance). These requirements may include the natural and induced environments in which the system will operate and any known constraints.

1. The dishwasher must be energy efficient.
2. It must connect to the standard water and power connections.
3. It should be easy to understand and have easy operator controls.
4. It needs to wash and dry kitchen items.
5. It must have different colored fronts to match different kitchen decors.
6. It must be quiet during operation.
7. It must be programmable so it can be run at some time in the future.
8. It should be easy to service.
9. It should hold the dishes, pots, pans, and utensils from a normal-sized meal.

Note the stakeholder requirements are usually very high level like the ones above. However, a stakeholder requirement can be as explicit as using a specific capacitor or screw if for some reason the stakeholder will not be satisfied with any other option. These types of low-level stakeholder requirements are often driven by compatibility, historical supplier issues/solutions, or many other legitimate reasons and should be vetted to ensure they are truly requirements. But the systems engineer needs to stay alert for wants of the stakeholder rather than true requirements. If the stakeholder asks for a superfluous capacitor, the systems engineer must negotiate for what is best overall. Adding an unnecessary feature can be done, but at an additional cost.

While the stakeholder may ask for too much, he or she may also ask for too little. The stakeholder requirements may imply more than they state. For example, requirement 4 states the system needs to wash and dry kitchen items. The systems engineer should determine that these kitchen items will be soiled by food, which may include grease, fat, and even possibly mold—we have all found that forgotten piece of cake in the back of our refrigerator. By requiring the system to wash the dishes, the stakeholder is really saying the dishes will go in dirty but must come out dry and sanitized so they can be used again safely.

Once the stakeholder requirements seem adequate, system requirements can be developed. While stakeholder requirements specify what the system must do, system requirements specify how the system shall accomplish those stakeholder requirements.

2.1.2 SYSTEM REQUIREMENTS

During the system architecture and requirements phase (Figure 2.9), the stakeholder requirements are analyzed, rewritten, and decomposed into well-formed system requirements. Through in-depth discussions and analysis of all stakeholder requirements, the systems engineer converts the stakeholder language into the

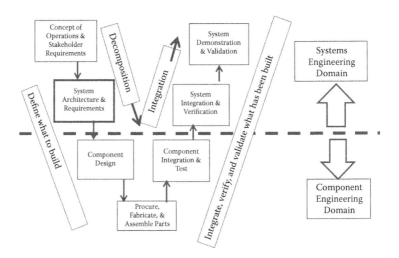

FIGURE 2.9 The second phase of the Vee.

system specification. At this level, requirements are categorized (e.g., functional, physical, performance, interface). Some of these requirements can move forward as written, while others are broken into more detailed requirements.

When writing a system requirement, it should be written as a well-formed requirement statement. That means it is in the following form:

Subject + Verb + Modifier

A properly structured system requirement will define the system of interest, the word *shall* (or in some places *must*), and then the description of what the system of interest shall accomplish. Let us note here that the stakeholder requirements will not usually be documented in this formal-looking format, but instead, they will be captured in the customer's voice. It is the systems engineer's job to translate them from the customer's voice to formal system requirements. Examples of system requirements for the dishwasher include:

- The dishwasher (subject) shall connect (verb) to a hot water supply (modifier). (Input)
- The dishwasher shall accept detergent. (Input)
- The dishwasher shall display an operational status indication. (Output)
- The dishwasher shall provide wastewater for disposal. (Output)
- The dishwasher shall accept a water temperature setting during maintenance operation. (Maintenance)
- The dishwasher shall output a diagnostic status during installation/maintenance operations. (Maintenance)
- The dishwasher shall fit into a rough opening of 35 7/16 in. (90 cm) height. (Installation)
- The dishwasher shall fit into a rough opening of 23 5/8 in. (60 cm) width. (Installation)
- The dishwasher shall fit into a rough opening of 22 7/16 in. (57 cm) depth. (Installation)
- The dishwasher shall be electrically grounded in compliance with the National Electrical Code, latest edition of ANSI/NFPA 70, in the United States or the Canadian Electrical Code, Part I, CSA Standard C22, in Canada or the local National Electrical Code. (Installation)
- The dishwasher water shall obtain a temperature of at least 140°F. (Operation)
- The dishwasher shall clean dirty articles in 40 minutes or less. (Operation)

All system requirements must be considered (Figure 2.10). A recommended approach is to consider all types of inputs and outputs for the system using an input/output matrix. A partial set of dishwasher input/output requirements is shown in Table 2.1.

The result of this effort is an approved, well-defined, controlled, and measured collection of baseline requirements and verification methods for a product, as well as a description of how the system will be used (concept of operations). We will explore the verification and validation of requirements in greater detail in Chapter 5, but for now suffice it to say that there needs to be a way to ensure every requirement is met.

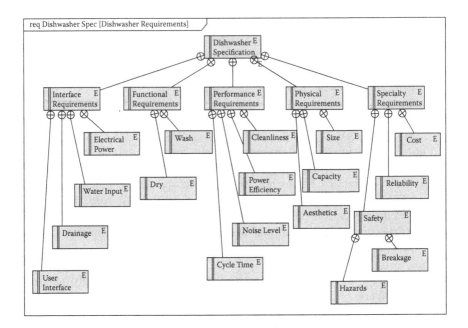

FIGURE 2.10 Example system requirements specification outline.

TABLE 2.1
Input/Output Matrix Example for Dishwasher

	Inputs		Outputs	
	Intended	**Unintended**	**Desired**	**Undesired**
Water	Clean and nominal pressure	High-pressure, rusty, dirty water	Gravity flow of dirty water	Overflow or underflow
Electrical	Nominal voltage	Surge voltages	Normal current	Electric shock
Dishes	Dirty kitchen items and utensils	Non-dishwasher-safe kitchen items	Sanitized kitchen items	Broken dishes
Detergent	Dishwasher detergent	Laundry or nondishwasher soap	Rinsed items	Suds, soapy film, and spots
Environmental	Normal household air		Room temperature, dehumidified air	Noise, vibrations, highly humid air

The requirements process is iterative, looping in reaction to design maturity, complexity, and change. The process is closed loop since the requirements are not satisfied or complete until the necessary verification methodologies are successfully executed against the design solution(s). A tracking method is a requirements verification traceability matrix (RVTM), which provides traceability from the requirement to the verification method in order to guarantee the requirement set is complete. The requirements allocation matrix should also provide traceability from the requirement to the system component so it can be quickly identified where in the system each requirement is being implemented. The RVTM also documents the associated verification method for each requirement; we will talk more about verification methods in Section 5.6.

2.1.3 SYSTEM ARCHITECTURE

Since many organizations struggle with the concept of architecture, we have broken it out into its own section. Yet we are continuing the discussion of the system architecture and requirements phase. In an attempt to avoid confusion, we will draw on the work performed by a building architect to explain this concept. When a new building is to be designed, the owners of the new building (stakeholders) meet with the architect. They discuss the location of the building, the size of the building, the number of floors the building is to have, the use of the building, and what kind of building style the owner prefers (modern, art deco, gothic, etc.). All of these details are what systems engineers call stakeholder requirements, as discussed in the concept of operations and stakeholder requirements phase. From those requirements, the architect begins the creation process, and the artifacts of that work are the architectural drawings. These drawings may depict what the building might look like, location plans, site plans, floor plans, and even some initial structural plans (Figure 2.11). These plans are shared with the new building owner, and changes are made.

Typical Civil Architecture Plans

- ❑ Presentation Drawings
  - » The sales pitch
- ❑ Survey Drawings
  - » Location
  - » Drainage
- ❑ Record Drawings
  - » As-built drawings - final versus sales pitch
- ❑ Working Drawings
  - » Foundation plan
  - » Framing plan
  - » Floor plans
  - » Sections and interior elevations
  - » Schedules for windows and doors
  - » Assembly drawings
  - » Component drawings
    - Windows, closets, cabinets, etc.
  - » Other trades: plumbing, electrical, etc.

Site Plan

Two-point perspective

Floor Plans

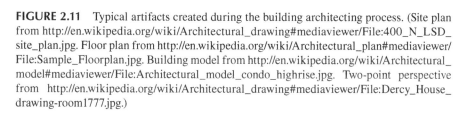

FIGURE 2.11 Typical artifacts created during the building architecting process. (Site plan from http://en.wikipedia.org/wiki/Architectural_drawing#mediaviewer/File:400_N_LSD_site_plan.jpg. Floor plan from http://en.wikipedia.org/wiki/Architectural_plan#mediaviewer/File:Sample_Floorplan.jpg. Building model from http://en.wikipedia.org/wiki/Architectural_model#mediaviewer/File:Architectural_model_condo_highrise.jpg. Two-point perspective from http://en.wikipedia.org/wiki/Architectural_drawing#mediaviewer/File:Dercy_House_drawing-room1777.jpg.)

If one pauses to consider that a building is a system, these plans represent the system architecture.

A typical top-level system architecture diagram may be a SysML block diagram representing the major subsystems of that system of interest. Other diagrams might show how the major parts interface with one another, how the system interfaces with its operational environment, and any other pertinent details of the system in its intended environment. The system architecting process begins by performing a functional analysis of the stakeholder requirements, which is discussed in greater detail in the next section. The functional analysis establishes the required functional behavior of the system. While there are a number of important tasks that a system architect should perform, Table 2.2 lists many of the significant tasks. While performing these tasks, the system architect will create a number of diagrams and models. A good starting point for those diagrams and models is shown in Table 2.3.

The system architecture and requirements phase demonstrates why systems engineering is an iterative process. To develop a good system architecture, the systems engineer must understand the system requirements. However, as the architecture is developed, more detailed system requirements will be discovered and others will be

---

TABLE 2.2

Typical System Architecting Activities

- Define a consistent logical architecture—capture the major logical elements and the interaction between them.
- Partition system requirements and allocate them to system elements and subsystems with associated performance requirements.
- Evaluate alternative design solutions using trade studies.
- Define the system integration strategy and plan (to include human system integration).
- Establish and maintain the traceability between requirements and system elements.
- Define verification and validation criteria for the system elements.

TABLE 2.3

Selected Artifacts Created during the Architecture Process

- Define a consistent logical architecture—capture the logical sequencing and interaction of system functions or logical elements.
- Partition system requirements and allocate them to system elements and subsystems with associated performance requirements—evaluate off-the-shelf solutions that already exist.
- Evaluate alternative design solutions using trade studies.
- Identify interfaces and interactions between system elements (including human elements of the system) and with external and enabling systems.
- Define the system integration strategy and plan (to include human system integration).
- Document and maintain the architectural design and relevant decisions made to reach agreement on the baseline design.
- Establish and maintain the traceability between requirements and system elements.
- Define verification and validation criteria for the system elements.

challenged to the real necessity. Therefore, these two tasks are usually done in short, iterative, refining cycles. It is not unusual for a systems engineer to be asked to add capability to an existing system. For instance, a manufacturer of heavy equipment may have a highly successful backhoe product, but it would be even better if the backhoe could also be fitted with a ditch-digging accessory. For the simplest extensions to an existing system, the systems engineer may treat the existing system as a black box, and then simply worry about how the new capability will interface with the existing black box. As the complexity of the additional capabilities increases, more advanced methods must be employed.

2.1.3.1   Functional Architecture

As the system progresses down the Vee's left side of the system life cycle (see Figure 2.9), the systems engineer determines what functions the system must perform. A function can be described as a discrete action (or series of actions) that is necessary to accomplish a specific objective or task. This function may be performed by one or more parts of the system and may be grouped logically by time, control, data, etc. The functional architecture therefore describes how a system is organized by what the system does. Of course, two different systems engineers may organize their functional architectures differently with similar yet different terms for the functions. In any case, this architecture defines what the system must do—the capabilities or services it will provide and the tasks it will perform. It also provides a description of the message/message type of communications between the functions, as well as the data that are passed between functions.

The functional architecture is created as the systems engineer analyzes the requirements and creates functions that might satisfy each requirement, or satisfy a collection of requirements. This functional analysis includes:

- A list of the core functions and function performance requirements that must be met in order to adequately accomplish the operation, support, test, and production requirements of the system, i.e., life cycle of the system
- A method for analyzing performance requirements and dividing them into discrete tasks or activities

While performing functional analysis and decomposition, the functions may be grouped by logical groupings, time ordering, data flow, control flow, or some other criterion. There are a number of tools and diagrams useful during functional analysis, but they are beyond the scope of this booklet. However, we will list a few so that you can look them up later if you desire. A partial list of tools and diagrams include the following (Figure 2.12):

- IDEF0 diagram
- Functional flow block diagram (FFBD)
- N^2 diagrams
- Timeline analysis
- Tree diagrams
- SysML (such as activity diagrams, sequence diagrams)

FIGURE 2.12   Engineer design. (From http://www.PresenterMedia.com/help.html.)

One common usage for functional decomposition is the IDEF0 (pronounced I-DEF-zero). IDEF0 is a compound acronym (ICAM Definition for Function Modeling, where ICAM is an acronym for integrated computer-aided manufacturing) associated with manufacturing functions. It offers a functional modeling language for the analysis, development, reengineering, and integration of information systems; business processes; or software engineering analysis (DAU 2001). In IDEF0, inputs always enter a function from the left, outputs always exit to the right, enablers enter from the bottom, and triggers enter from the top (Figure 2.13). The function in the box represents some transformation.

For the dishwasher, the top-level function may be to "clean dirty kitchen items." This single function shown in Figure 2.14 can now be broken into smaller subfunctions (decomposed). If the main purpose, or function, of a dishwasher is to clean dirty kitchen items, then the dishwasher is the resource, or enabler, for the function. As we said, the enablers are represented by an arrow entering the function from the bottom. Required inputs, represented by arrows entering from the left, are dirty kitchen items, water, soap, etc. Examples of what this function outputs include clean kitchen items, dirty water, and heat. Finally, commands are the trigger that initiate this function, and therefore come in from the top.

Using the IDEF0, the next task is to determine what collection of subfunctions could satisfy this function. The totality of these subfunctions must encompass all that is implied by the level above, in this case, "clean dirty kitchen items." To complete Figure 2.15, the systems engineer must identify the necessary inputs, outputs, resources, and triggers that would allow the subfunctions to work with one another to satisfy the system mission.

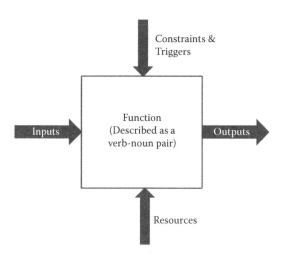

FIGURE 2.13   The IDEF0 diagram.

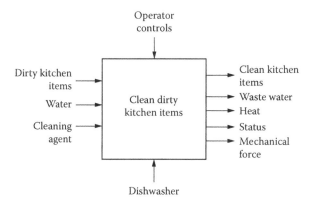

FIGURE 2.14   Top-level function for a dishwasher.

A final note: When working on the functional (or logical, as it is sometimes called) architecture, try not to represent functions with physical names. For instance, do not talk about a valve; instead, maybe call it a fluid control device. Physical names will cause confusion when you transition from the logical architecture to the physical architecture, which will have physical names.

2.1.3.2   Physical Architecture

The physical system architecture is a collection of graphical depictions of the system of interest under consideration. The physical architecture defines the partitioning of the system resources (hardware and software) needed to perform the functions. It should also show the interconnections between other systems, and between the subsystems. There should be a diagram representing the functional architecture allocated to the physical entities, since every item performs some function for the system.

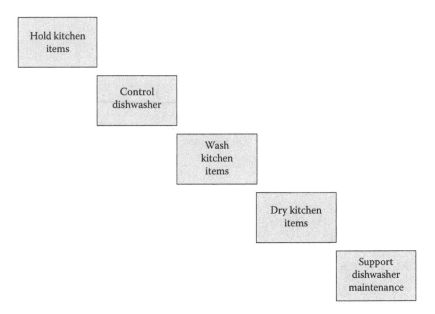

FIGURE 2.15   Example of possible dishwasher subfunctions.

2.1.4   COMPONENT DESIGN

The component design phase begins when formal requirements have been developed and the system architecture is defined (Figure 2.16). From the physical architecture, the subsystems have been identified, and the system requirements have been allocated to those subsystems. Formal requirements are the source of system verification, which will be discussed later. While system verification to requirements is definitely warranted, this phase may or may not include system validation. Nonetheless, this phase is usually long, and it is the main phase for all the traditional engineering disciplines. Component design ends when the system is ready for production.

During the component design phase the systems engineer will be expected to support the entire engineering team. He or she should continue to be the visionary for the overall system during the ongoing engineering process. Consequently, the design phase requires the systems engineer to work closely with and respect all engineering disciplines on the design team. In addition, the systems engineer will be required to make interface decisions and to resolve trade-offs between different subsystems of the overall system (Figure 2.17). As always, the systems engineer should remain cautious of becoming arrogant regarding the system, but instead rely on the whole team of engineers. Working together and listening to each other, the entire team can make sure the decisions, risk, and opportunities that were determined in the conception phase are holding true.

A recurring theme, the systems engineer should ensure the system design is not straying from the original system concept and architecture. Many times during the system design the system concept is, for lack of better words, "thrown out the window." Unfortunately, many systems fail because they do not meet the original needs

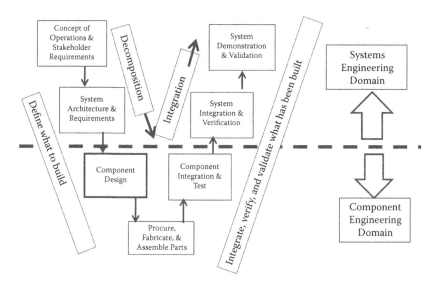

FIGURE 2.16   Third phase of the Vee.

FIGURE 2.17   Communication is key.

of the stakeholders. Validation attempts to prevent this failure, and we will discuss validation later in this chapter and more thoroughly in Section 5.7.

2.1.5   PROCURE, FABRICATE, AND ASSEMBLE PARTS

The procure, fabricate, and assemble parts phase occurs partly in parallel with component design. As the design is developed, the thought of how the part is going to be procured, fabricated, and assembled should be part of the design discussions. It would be useless to design a part that cannot be procured, fabricated, or assembled into the system or that is so expensive it does not meet the cost goals of the product. Many engineers have the attitude that anything can be done, which has some truth to it, but that attitude must be managed within the stakeholder requirements.

At the end of the procure, fabricate, and assemble parts step, the components should be ready for integration with other parts and ready for the verification process.

2.1.6   COMPONENT INTEGRATION AND TEST

The component integration and test phase begins once the components have been fabricated and assembled (Figure 2.18). Components are integrated with one another until the complete system is assembled.

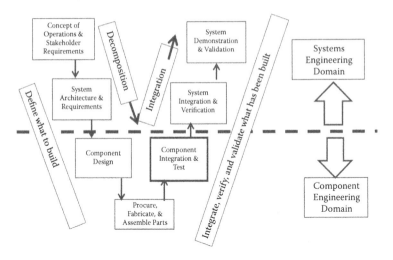

FIGURE 2.18 Fifth phase of the Vee.

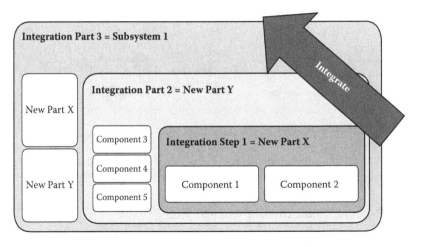

FIGURE 2.19 Example integration plan.

Prior to this step, an integration plan should be developed to describe when and how each component will be integrated. There should also be a verification plan to test the components as they are integrated with one another. A good integration plan calls for testing during integration and not just the integration of all the components at once before beginning the verification process (Figure 2.19). It is often very difficult to determine the source of issues once component integration is complete.

2.1.7 System Integration and Verification

In any systems engineering effort, there are system physical components (such as hardware items, software items, constituent systems within a system of systems, or the human systems that interact with the system) and functional/logical components

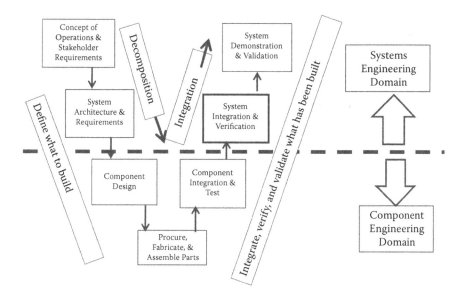

FIGURE 2.20   Sixth phase of the Vee.

(such as algorithms and system processes, operator processes, or business processes). Both sets of components must work together for the entire system to succeed. System integration and verification is the phase that emphasizes the assembly of parts into a system that meets the system requirements (Figure 2.20). System integration unifies the system physical components and the functional/logical components into a whole (Figure 2.21). It ensures that the hardware, software, and human system components interact to achieve the system purpose and satisfy the customer's need.

Verification confirms the system of interest and its elements meet the specified system requirements. In other words, it asks the question: Was the system built right? The validation step will then answer the question if the right system was built.

2.1.8   System Demonstration and Validation

Validation confirms the completed system in its intended environment satisfies or will satisfy the stakeholders' needs. It determines a system does everything it should and nothing it should not do. In other words, it asks the question: Did we build the right system?

Although represented only on the right side of the Vee model (Figure 2.22), there are two important periods of time when validation should be performed. Product validation confirms the completed system meets its intended purpose. Early validation ensures the stakeholder requirements and concept of operations meet the needs of the stakeholder. Often the stakeholders express what they want rather than what they need. Even more likely, the stakeholders are not able to effectively communicate what they need. Early validation mitigates this problem and attempts to ensure the true need is captured.

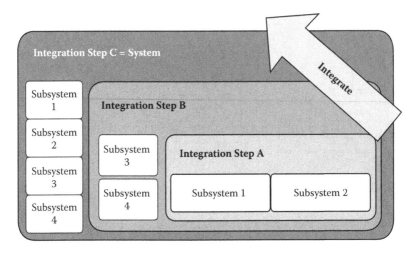

FIGURE 2.21   System integration example.

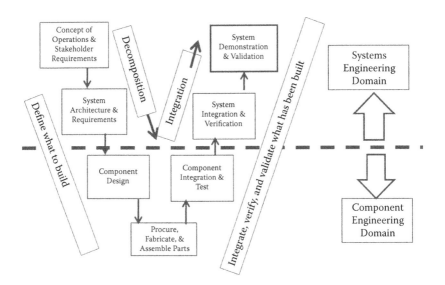

FIGURE 2.22   Seventh phase of the Vee.

2.2   SYSTEM PRODUCTION

There are a few additional phases after the system has gone through the Vee model of design and build. The system production phase begins once the system has been validated and handed off to manufacturing for production of the product. The systems engineering responsibility during this phase is a support and oversight role. The systems engineer should have worked with manufacturing during the design to guarantee the product is designed for manufacturability. During manufacturing, the systems engineer should continue to make sure the product is manufactured as desired and ensure that liberties on changing the product (i.e., part changes, etc.) are not taken by the manufacturing team. The systems engineer should also be a reviewer of any production test plans. Keeping the systems engineer in the loop during this time can avoid issues that arise from the final product not meeting its intentions.

2.3   SYSTEM UTILIZATION AND SUPPORT

The system utilization phase begins once the production level product is fielded to the end user. Usually by this phase the systems engineer becomes a monitor of the product. During this phase field engineers and sales representatives are the closest people to the product. The systems engineer should be seen as a shepherd of the product and kept in the loop since he or she will have invaluable knowledge of the product, allowing him or her to be helpful in the debugging of any system issues. Also keeping the systems engineer linked to the product in the field can be invaluable in learning lessons for the next generation of the product.

During this phase is when the true validation of the product can happen since the system will have its first real chance to meet its need. While validation is attempted during design to decrease the risk of product failure, the most complete validation can only be done when the manufactured final product is fielded with end users. This final validation is the ultimate stamp if the right system was built. Accordingly, the systems engineer should be involved in any feedback from the end user.

Once the product has been in the field and utilized, the engineering support for the system will decrease, but the systems engineer should stay involved in issues as they are found. During this support role the systems engineer could have invaluable solutions and knowledge to support field engineers, sales, and customers.

2.4   SYSTEM RETIREMENT AND DISPOSAL

The system retirement and disposal phase begins when the system is ultimately taken out of service until the time the system is properly disposed. The systems engineer still serves a support role and helps assure the system is properly retired and disposed. A good system design should have included plans for disposal and retirement, and the systems engineer should be available to answer any questions related to these plans, such as the intent and process of the disposal. If the system is part of another system, the retirement might require specific processes to avoid impacting the other system, and the systems engineer should be available to answer questions on how the retirement was designed.

3 Other Systems Engineering Development Models

The Vee development model has been used to discuss the various general tasks performed by systems engineers. It is now time to address a few of the other development models that are in use today. The first one that will be discussed is the spiral model, which was created for software development in the late 1980s but is currently widely used in system development. The second one is a more recently developed model that uses an agile process.

3.1 SPIRAL MODEL

The spiral model is an iterative process that was first developed for software development. The iterative nature of the process makes it ideal for large system development. The process starts in the center of the determine objectives, alternatives, and constraints quadrant of the spiral in Figure 3.1, and then moves clockwise and outward. There are four main phases of each cycle through the process: (1) determine objectives, alternatives, and constraints; (2) evaluate alternatives—identify and resolve risks; (3) develop and verify next level project; and (4) plan next phase. One of the benefits of the spiral model is that before the start of a new cycle there is a review and commitment to move forward.

The spiral model (Figure 3.1) incorporates the waterfall model, which is an older, sequential design process. The waterfall model has fallen out of favor since it assumes every phase is fully complete and correct before moving on. Even under ideal conditions, new information may become available between the initial systems concept and producing it. Therefore, the various system models now incorporate iterations, although in different ways. The spiral model is an iterative process that also adds risk analysis, prototyping, and planning. It is also better for larger system development since pieces (subsystem, component, etc.) of the system that are higher risk can be developed earlier in the process and then assessed before moving forward to the next iteration.

3.2 AGILE MODEL FOR SYSTEMS ENGINEERING

The agile model known as scrum is shown in Figure 3.2. The model is performed in an iterative nature with feedback from the stakeholders after an iteration of the process model. One of the main properties with the agile method is the concept of

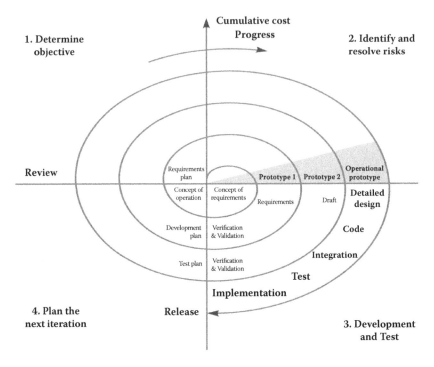

FIGURE 3.1 Spiral development model. (From http://commons.wikimedia.org/wiki/
File:Spiral_model_%28Boehm,_1988%29.svg.)

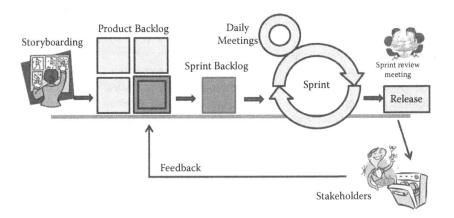

FIGURE 3.2 Agile scrum process.

sprints, which are set time limits usually in the time frame of a few weeks. This idea
of setting time limits and pushing through the process helps to assure the progression
of the system development, which in other models can become stalled.

The first step in the agile scrum model is storyboarding. The idea of storyboarding is similar to use case development, but the storyboard is usually more visual

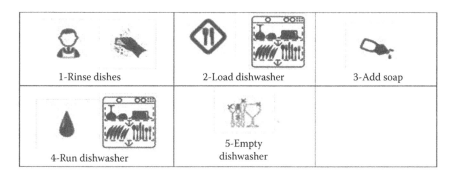

FIGURE 3.3   Storyboard showing how to use a dishwasher.

(versus the textual nature of use cases). Keeping with our example of a dishwashing machine, an example of storyboarding is shown in Figure 3.3. From the example, one can see how a system will be utilized over time. Like use cases and operational concepts, there will be more than one storyboard for a system, but each sprint cycle of the process may focus on only one storyboard.

The second step is the product backlog, which is similar to the requirements elicitation discussed in Chapter 2. The system requirements are developed in this step. Then a portion of the requirements or product backlog is selected for the sprint cycle, which is called the sprint backlog. Once chosen, the sprint begins. The sprint is the design cycle and contains daily meetings to discuss the current status of the sprint. At the end of the sprint, there is a review of only the finished work (or release) with the stakeholders. Feedback from the stakeholders is incorporated back into the product backlog, and the process repeats until all the product backlog has been addressed.

There are many other systems engineering development models in use today. Although we have only discussed a few of them, we hope these examples give an idea of why different ones are needed depending on the situation. We encourage the reader to find out what development model is in use where he or she works or is interested in working.

4 System of Interest

Some argue any system may be decomposed into multiple systems and any system of systems may integrate into a higher-level system. Although we choose more specific definitions for systems, as opposed to parts, and system of systems, as opposed to traditional systems, one person's system may be another person's subsystem. A level of abstraction indicates the system of interest for each perspective. Accordingly, one person's system of interest may compose or be a component of another person's system of interest.

4.1 ABSTRACTION AND DECOMPOSITION

When developing or discussing a system in any context (architecture, requirement, design, etc.), one must be aware of the level of abstraction that is being discussed. A system can be discussed in context anywhere between a system of systems and the component/element level. Determining these levels of abstraction for a system can sometimes be challenging but should be clearly defined so that everyone can use the same language regarding a particular instantiation of a system. For example, the air traffic control system contains many interacting systems. So it can be viewed as a system of systems, although the air traffic control system is a system within the overall transportation system (i.e., trains, planes, automobiles, etc.). This example indicates the importance of understanding the context in which a system is being designed and discussed. Just as an individual may participate as the role of mother, teacher, soldier, and coach, a single system can play multiple roles, such as system, module, and so forth. The system must be designed to fulfill all its roles, and knowing what the scope of the system is contributes to that goal. It is important to know the boundaries of the system for understanding what is within a system and what is outside the system. Therefore, one of the first steps when designing a system is to decide what level of abstraction applies to the system of interest.

A definition related to abstraction provided by Shaw (1984) is "a simplified description, or specification, of a system that emphasizes some of the system's details or properties while suppressing others. A good abstraction is one that emphasizes details that are significant to the reader or user and suppresses details that are, at least for the moment, immaterial or diversionary."

To determine the appropriate level of abstraction, one must understand the different hierarchy levels of a system, which are shown in Figure 4.1. On first thought, the hierarchy level of a system may appear to be easily labeled, but as Booch et al. (2007) pointed out, "abstraction focuses on the essential characteristics of some object, relative to the perspective of the viewer." Therefore, depending on a person's perspective, the abstraction level (system, subsystem, etc.) may be classified differently.

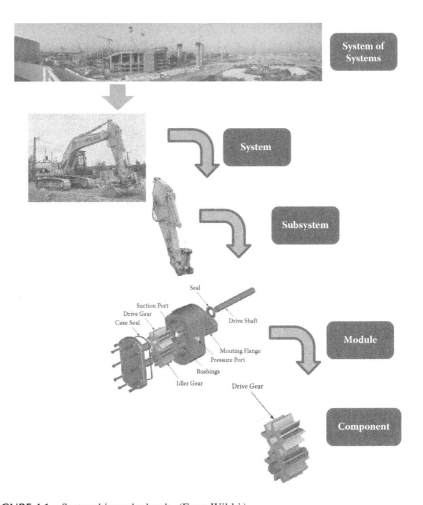

FIGURE 4.1 System hierarchy levels. (From Wikki.)

There are some questions one can ask to help determine the level of abstraction. Some of these questions are as follows:

1. Who will be using the system today and in the future? In other words, from whose perspective should the system be developed?
2. What parts of the architecture are important to those who will be utilizing the system today and in the future?
3. What is difficult to understand about the system?
4. What parts of the architecture will be common or used over and over again?
5. How will the system be used?
6. What is the new technology in the system?
7. Are there parts of the system that may be confusing? Were there parts of the system that people (engineers, customers, management, etc.) asked more questions about than others?

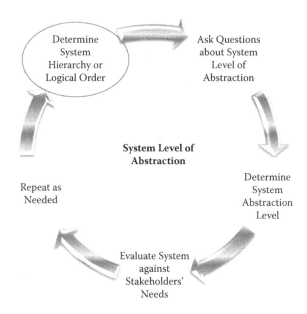

FIGURE 4.2 Abstraction process.

The process for determining the level of abstraction might look like Figure 4.2. From Figure 4.2, one should start at the step of determining the hierarchy or logical order for the overall product or project. Next, one would ask the aforementioned questions in order to determine the level of abstraction. Once the abstraction level is known, then one should evaluate the abstraction level with all stakeholders, which determines if it is at the appropriate level. This process is repeated until stakeholders are satisfied that the appropriate abstraction level has been defined.

4.2 INTEGRATION

Integration is a process that unifies the system components and the process components into a whole. Referring back to Figure 4.1, it is a way to combine the components into modules, the modules into subsystems, the subsystems into systems, and potentially the systems into a system of systems. In other words, it is a way to combine elements at one level of abstraction together in such a way to fulfill the role of the next higher level of abstraction. There are system components (such as hardware items, software items, constituent systems within a system of systems, or the human systems that interact with the system) and there are process components (such as algorithms and system processes, operator processes, or business processes). Both sets of components must work together for the entire system to succeed. Different techniques may be needed at each level of the hierarchy, and integration activities happen as many times as needed. The process of integration must ensure any hardware, software, and human system components will interact to achieve the system purpose and satisfy the customer's need. By principle, the integration process should improve performance, reliability, and interoperability without adversely affecting existing or proposed system functionality and operations. Of course, it is not as easy as it sounds. For one thing, integrating elements of a system can have a direct impact on the interfaces, both the internal and external, including the human-system interfaces.

Despite its difficulties, integration is vital for the assembly of a system, regardless of complexity, that meets the needs of its intended users. The activities performed by systems engineers ensure that combining the lower-level elements results in a functioning and unified higher-level capability.

In Chapter 5 we discuss the development of requirements in more detail, but we cannot discuss integration without mentioning the role of requirements. First and foremost, the requirements are used to develop a plan for integration. When the systems are assembled, they must meet the requirements; otherwise, the integration would not be a success. Furthermore, the systems engineer must stay mindful of existing legacy requirements, as well as new governance and logistical attributes, which may affect how the system is maintained or supported. Finally, the systems engineer must be aware that integrated systems may behave differently than expected from the components' behaviors. Requirements that appeared to have been met may fail after integration, yet other requirements may be met. The systems engineer needs to consider how to integrate the system such that the desired capabilities are met.

Although we only briefly discuss integration in this section, its importance should not be underestimated. Sloppy integration will result in a system that does not meet its purpose, and the integration activities affect many aspects of engineering a system, including requirements management, the conceptual design from stakeholder requirements down to the specifications, and various levels of architectures, testing, and production.

5 Developing and Managing Requirements

Developing and managing requirements through the evolution of system development have been identified as one of the top factors for overall success of the system (Charette 2005; McManus and Wood-Harper 2007; Christel and Kang 1992; Standish Group 1994). Requirements are in essence the foundation of the system that is being built and also the foundation for the project (cost, schedule, etc.). Therefore, this section will be dedicated to discussing how requirements can be developed and managed.

At first, it may be easy to think requirements development is a clean-cut process. A customer or marketing team, depending on the industry, approaches the designers with a completed set of requirements that have been polished to perfection and are ready to move into the design phase. Although this situation would be ideal, it is of such extreme rarity that some may argue it never happens. In fact, some customer requirements are not solidified until well into the design and development of the system. This fact haunts many engineers since the requirements are the foundation. It can feel as if a system is being built on shifting sand, and any engineer may get frustrated trying to build the best system upon an ever-changing foundation. Therefore, the systems engineers must manage the requirements in such a way that the requirements can change with minimal impact to the system design. Managed requirements allow all the other disciplines working on the project to proceed forward with a stable foundation.

5.1 CYCLONE REQUIREMENTS MANAGEMENT PROCESS

The cyclone process* is a requirements management process that combines the spiral model and the inquiry-based requirements model (Zhang and Eberlein 2002) with a third dimension added. As the name implies, this model can be seen as a cyclone. From the top 2-D view (Figure 5.1a), an iteration of the spiral has four phases: (1) requirements elicitation, (2) requirement discussion and analysis, (3) requirement negotiation and commitment, and (4) requirement specification. The whole spiral has a third dimension, which is the risk dimension (Figure 5.1b). The risk element sets this process apart from other requirements processes. The process is performed in multiple iterations with the requirement-associated risks decreasing. The ideal goal

---

* The cyclone requirements management process was first proposed in a master's thesis by Mary Bone and published as Cyclone Process: Dealing with Vague Requirements, in *INCOSE Conference on Systems Engineering Research*, Los Angeles, 2008.

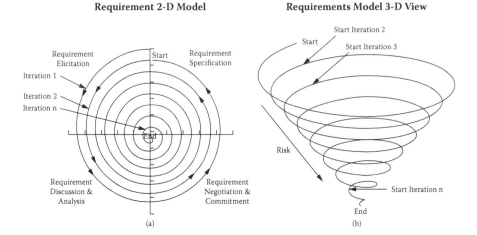

FIGURE 5.1 Cyclone requirements process model, 2-D and 3-D views.

would be to eliminate risk totally, but realistically the goal should be to minimize risk to an acceptable level agreed upon by all stakeholders.

So the main idea behind the cyclone process is managing risk of changing requirements through a well-documented cooperative approach. This process allows for poorly defined requirements at the onset of a system project to exist, and for the requirements risk to be mitigated by other well-defined or additional requirements. For example, if it is known that a product will be painted but the type of paint cannot be decided up front, then additional requirements may be defined for the product to mitigate the unknown paint requirement. The mitigation requirements may attempt to define the surface that can be painted by most of the known paints used in that industry. Then, when paint is finally chosen, the risk of it not working with the chosen surface is very small. It is important to record the rationale behind each mitigation requirement. Continuing the example, the surface requirement should be labeled as a mitigation requirement due to the inability to fully define the paint requirement and indicate it was not derived from the stakeholder. Also, there should be a paint requirement that is stated as a TBD (to be determined), which traces to the other defined requirements. Due to the complexity of requirements and their relationships, five pieces of information must be captured for each requirement in the cyclone method: (1) requirement statement, (2) requirement rationale, (3) requirement relationships (traceability), (4) requirement risk factor, and (5) requirement risk factor rationale.

The cyclone process breaks from the traditional requirements process in that the goal is not to fully define the requirements, but to define the requirements to a level that has an acceptable risk factor to move the system development forward. When a set or subset(s) of requirements is defined to an acceptable risk factor, they are allowed to continue through the system process. Addressing subsets (i.e., for a subsystem) of requirements allows those subsets to proceed while other subsets go through more iterations of the requirements process.

5.2   REQUIREMENTS ELICITATION

Requirements elicitation is an activity to obtain the requirements for a system. In the cyclone requirements process model, potential, fully defined requirements should be captured along with requirements that cannot be fully defined (according to current requirement standards). The systems engineer's role is to determine all he or she can about the potential requirements and define the requirements as thoroughly as possible. The systems engineer must also capture rationale and justification for each requirement.

For example, during this activity, the customer states he or she cannot disclose where the product will be used. Accordingly, the systems engineer must determine a best-fit requirement for the countries in which they believe the product will be used. Along with the requirement, there must be a description stating the customer cannot disclose in which countries the product will be used and an explanation of how the potential countries were chosen. All information should be captured for the requirements so that no information is lost. This step is very important. For example, the requirement could be something like "The product shall be used in the following countries: Germany, France, Iraq, Afghanistan, Great Britain, and the United States." Along with this requirement, it should be explicitly stated that the customer/ stakeholder was not able to provide the list of countries and the systems engineer derived the list from obtained information and research. The research and references should also be captured with the requirement for future use.

During the requirements elicitation activity, risk is managed by documenting as much information as possible about each requirement. This is done so that an initial risk assessment can be performed in the next activity, requirement discussion, and analysis. Information for each requirement should include: (1) source of requirement, (2) known issues with requirement, (3) rationale for requirement, (4) relationship of requirements (how does this requirement affect other requirements), and (5) scenario development (e.g., use cases).

5.3   REQUIREMENT DISCUSSION AND ANALYSIS

Once the initial activity of elicitation is completed, the next task is for *all* the stakeholders to discuss and analyze the requirements gathered, along with the information for each requirement. Each requirement is analyzed individually. Prototyping or research may be required to better define poorly understood requirements and those needing more development. Any information discovered or created is captured along with the requirement. Also, any discussion points brought up by stakeholders are captured for future reference since it is not possible to remember every discussion or decision (even the important ones). The result is a set of requirements with accompanying analysis, information, and any discussion notes available for future reference.

Another task at this point can be allocating the requirements into subsets. The subsets are determined by the information gathered in the elicitation phase. Subsets of requirements can be chosen in many ways but are usually chosen due to their relationships with one another. In smaller systems, subsets may not be possible. However, in larger systems, there usually are clear subsets of requirements. In either case, the set or subsets of requirements as a whole should be checked for completeness, ensuring all requirements are allocated. The rationale for the allocation of specific subsets should be captured.

It should be obvious by now that the rationale and captured information for each requirement should be reviewed and agreed upon. For example, suppose stakeholder Bob Smith is stated as the source for a requirement. If he does not recall being the source of knowledge for the requirement, then there may be an issue with the requirement.

With the agreed set of requirements, risk assessment is performed. The systems engineer analyzes each individual requirement and assigns a risk factor. The risk factor should be based on:

- Where did the requirement originate?
  - A previously proven reliable source would lead to a low risk.
  - An unknown or new source could lead to a higher risk.
- How likely is this requirement to change?
  - Is there any current knowledge that makes one believe this requirement may change (i.e., regulation changes are being put through as these requirements are being developed that could impact this requirement)?
- Does the requirement meet all the basic criteria of a good requirement (see Alexander and Stevens (2002) or Blanchard and Fabrycky (1998) for criteria)?
- What is the impact of changing this requirement?
  - Even when the requirement is believed to be very stable, if changing it could cause a large impact, it could be a high-risk requirement.
- How many other requirements will change if this requirement changes?
- How does this requirement impact potential design?
  - This applies mainly when dealing with legacy systems that already have a set design or when dealing with a new system that has many constraints.
- Has the requirement already changed multiple times during the elicitation and early discussion and analysis phase?
- Does this type of requirement have a history of causing change or being changed?
- Who has been involved in the requirements process so far?
  - If most stakeholders have not been actively involved in the requirements process, then most requirements may be high risk.
  - If most stakeholders have been involved in the requirements process, the requirements may be at lower risk.

The risk factor scheme can be chosen by the user of the process. For some, high, medium, and low risk factors may work, while others may choose a 1–10 risk factor scheme. Either way is all right as long as the scheme is understood and defined. For example, if a risk factor scale of 1–10 is used, then make sure to define that 1 means it is low risk and 10 is high risk, or vice versa.

The assignment of the risk factor is essential in this process. It is likely there will be many requirements or subsets of requirements that can be defined on the initial iteration. Yet, other requirements or subsets may take much iteration to get to a point where they can be useful to the customer.

Again, the rationale for the chosen risk factor should be captured. This could be as simple as stating the requirement is a legacy requirement that has been solid for the last 20 years, which justifies a low risk. However, the rationale may be more complex. The risk factor may be derived by computing the risk based on lack of information regarding the topic (which increases risk) and mitigation requirements (which decreases risk).

Along with individual requirements, the whole set and subsets of requirements should have an overall risk factor assigned. If all the individual requirements in a subset have a low risk factor, then most likely the subset of requirements overall would have a low risk factor. This may not be the case, though, if there is believed to be a large number of missing requirements from the subset. Then the subset may have a high or medium risk. Whatever risk factor is assigned, it should be justified with documented rationale. This could be as simple as saying, "This set is believed to be lacking many requirements and therefore is high risk."

5.4   REQUIREMENT NEGOTIATION AND COMMITMENT

In this activity the stakeholders negotiate any changes and commit to the set of requirements for moving forward. The stakeholders review the requirements once again, but this time they look also at the risk factor assigned to each requirement. The stakeholders are able to sign up for the requirements as they exist in the previous activity, or they can negotiate changes. In fact, the risk factor can be used during negotiation. If the requested change actually causes a decrease in the level of risk, then that could be a valid reason for the change. Any change to the requirements should be documented along with the rationale for the change.

For each requirement there is negotiation and commitment to the risk factors assigned during the discussion and analysis activity. It is very important that all the stakeholders agree, not only to the requirement, but also to the associated risk. The commitment to the risk is just as important as commitment to the requirement itself. If a stakeholder believes that a requirement is a high risk but another believes it is a low risk, it is important that they discuss and negotiate a risk factor that is acceptable. It could be that one knows something the other does not and maybe an update to the risk rationale is all that is needed.

The same argument applies to the negotiation and commitment to the set or sets of requirements also. An agreed low-risk set of requirements could be ready to go to the next phase of the overall system process, while those with high-risk factors may need to proceed through another iteration of the requirements process. However, some projects may be willing to go forward with a few high-risk requirements. Different projects are willing to accept different levels of risk.

5.5   REQUIREMENT SPECIFICATION

In this activity the requirement specification is written and reviewed. Not only is the requirement captured in the requirement specification, but also information is gathered through all the previous phases. If the requirements have been divided into subsets, then each subset should be clearly defined in the requirement specification.

The team identifies requirements that need to be put through another iteration of the process based on documented information and the assigned risk factor. Attributes are assigned to requirements needing another pass of iteration. This allows *all* project members to view the requirement but understand additional work is needed.

The risk factors are documented in the requirement specification. A very volatile (high-risk) requirement can be captured in the requirement specification, but the associated risk factor should be clearly displayed. This allows customers of the requirement specification (usually design engineers or subsystems engineers) to understand the state of the requirement while getting to see a complete picture of the current system.

The risk factor assigned to the overall set or subsets should also be clearly stated. This allows for some subsets to possibly proceed through the system process while other subsets of requirements go through another iteration of the requirements process.

5.5.1 REQUIREMENTS DECOMPOSITION

As requirements are generated through the requirements process, they are decomposed into more and more specific requirements. Derived requirements are identified that are necessary to satisfy the initial stakeholder/customer requirement. The decomposed requirements are usually allocated to the subsystem/component teams that are functionally associated with the given requirement. One of the hardest tasks of decomposing requirements is distributing the requirement across the lower-level tiers (subsystems, components, etc.). For example, the cost of a system is usually a stakeholder requirement. The cost has to be distributed across the next lower level, etc. Deciding on the distribution and which subsystem, component, etc. gets how much of the cost distribution can require much effort and negotiations among the engineering and management teams. Therefore, the subsystem teams work closely with systems engineers to determine the allocation and to perform a detailed decomposition/distribution of the requirements. These decomposed and derived requirements are added to the requirements management database, where they continue to evolve through tightly controlled processes for the life of the program.

The requirements allocation produces a flow of requirements from the system level down to the lowest implemented component and allocated to specific system functions, hardware elements, and software elements (Figure 5.2).

The allocation of requirements forms tiers of lower-level requirement specifications. These allocations from one level to the next are captured in a requirements allocation matrix. Design documentation captures the implementation that meets requirement specifications at each level. The requirements of lower-level specifications are usually traced to the next higher-level requirement specification. Tracing requirements assures that all stakeholder requirements have been implemented into the lowest-level design and allows insight into which subsystem(s), component(s), etc. fulfill a specific stakeholder requirement.

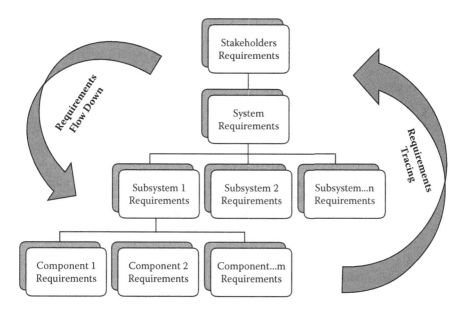

FIGURE 5.2 Requirements hierarchy.

5.5.2 Requirements Management

As already stated, the requirements process requires many pieces of information to be captured along with the requirement. There are a variety of software tools to help manage requirements and associated information, making the requirements process more manageable. These tools can help manage the requirements, but they are only tools. The systems engineers and the project team must assure the requirements and associated information (including tracing) are kept up to date.

At a determined point in time, each requirement along with associated information should be put under a type of change management system. This system prevents changes without approval and acknowledgment from those using the requirements data. A goal of requirements management is to ingrain into the culture of systems engineers the continuous reviewing and monitoring of the requirements and associated information (rationale, tracing, etc.) for correctness based on the current design of the system, and vice versa. With that goal in mind, the fundamental principles for requirements management can be stated as follows: (1) capture each requirement and all associated information, (2) formally review the requirement and associated information, (3) place the requirement and information under change control management, (4) periodically audit the requirement and information against the current design for correctness, and (5) update information as needed through the change control process.

As an ending thought, many times there is extensive effort to capture the right requirements only to have them set aside in favor of newer requirements that may or may not meet the initial needs. This practice is dangerous and should be avoided.

Everyone working on the system needs to keep asking, "Does this meet the top-level stakeholder and system requirements?" rather than just focusing on meeting the more familiar subsystem or component-level requirement. To develop the best systems, a culture should be generated that holds everyone on the project responsible for adhering to stakeholder and system requirements.

5.6   VERIFICATION

Now that the process for capturing the requirements and associated information has been completed, it is necessary to turn to the activities of validation and verification. Outside of systems engineering, the terms *validation* and *verification* are often used synonymously. Systems engineering gives them separate meanings, but even within the systems engineering community there can be some confusion over the differences. At a minimum, systems engineering defines verification as "building the system right" and validation as "building the right system." What this means is that verification is the process of making sure the system meets the system requirements (via testing, analysis, demonstration, or inspection), and validation is the process of making sure the system that will be or was built is the one the stakeholders need or want.

In the next section, we will discuss validation, which starts early in development with early validation and ends with a completed system at end product validation. Since the Vee model indicates verification before validation, we will discuss it before going into detail about validation. However, this representation is a shortcoming of the model since validation and verification occur throughout the system development, with emphasis on validation at both ends of the Vee model and verification emphasized at the bottom (Figure 5.3). Verification and validation are equally important because validation skips the necessary checks that go on in the middle phases of development. Verification fills this void by ensuring the system is developed according to the requirements, which is another reason the requirements are so important. Another way to think about it is verification determines the system is built as specified. Since the requirements should trace to validated documents, in theory the system should work as expected. But engineers, being people, may miss a requirement or misunderstand a stakeholder requirement. Over the life cycle of a system, many things can happen so that a system that meets system requirements does

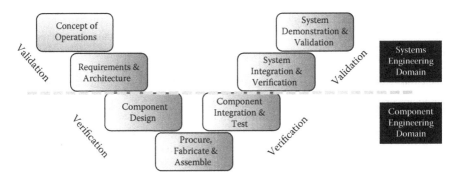

FIGURE 5.3   Validation and verification in the Vee model.

not fulfill the stakeholder requirements, which is why validation is also essential before a system is finished.

A verification plan should be developed that outlines the method that will be used to verify *each* requirement along with the pass/fail criteria. In addition to its use as a traceability tool, as discussed in Section 2.1.2, the requirements verification traceability matrix (RVTM) documents these verification methods for each requirement.

So we are talking about verifying the system against the requirements, but what do we mean by verifying? There are four main types of verification methods (Mil-Std 961E 2003), as follows:

1. Test: The use of scientific principles and procedures to determine the properties or functional capabilities of items.
2. Demonstration: The actual operation of an item to provide evidence that the required functions were accomplished under specific scenarios. The items may be instrumented and performance monitored.
3. Analysis: The use of established technical or mathematical models or simulations, algorithms, charts, graphs, circuit diagrams, or other scientific principles and procedures to provide evidence that stated requirements were met.
4. Inspection (or examination): A generally nondestructive activity that typically includes the use of sight, hearing, smell, touch, and taste; simple physical manipulation; and mechanical and electrical gauging and measurement. The item or system does not need to operate in order for this verification method to occur.

To put these definitions in clear language, inspection looks at a system that is in an unchanging state. Back in Section 2.1.2, several requirements were identified for a dishwasher. An installation requirement was "The dishwasher shall fit into a rough opening of 35 7/16 in. (90 cm) height." The system is verified against this requirement by inspection in that the systems engineer can look at the dishwasher and see it is the correct size. Demonstration is the observation or measurements when a system functions, or changes states. An input requirement was "The dishwasher shall accept detergent." The systems engineer can load the dishwasher with detergent and see that it accepts it. This demonstrates a normal state change of the dishwasher, as it starts out empty and becomes loaded with detergent. When a test is performed, the system state is manipulated and the results measured. There is a maintenance requirement: "The dishwasher shall output a diagnostic status during installation/maintenance operations." A test would be to artificially introduce different, maybe even extreme or fault, states and make sure the diagnostic status is appropriate. Finally, analysis is conducted using information known about the system or its elements without the system physically present. An operation requirement is "The dishwasher shall clean dirty articles in 40 minutes or less." If it has been verified that the wash cycle takes 15 minutes, the rinse cycle takes 10 minutes, and the drying cycle takes 15 minutes, then analytically the system is verified against the requirement.

Once a verification plan has been developed, it should be reviewed and agreed upon by the different component engineers. Unlike validation, verification can be performed without the stakeholder. Actually, the stakeholder may not understand the detailed nature of verification. Remember, the stakeholders are not the technical experts, which is one reason the systems engineer is so important. Once the verification plan is agreed upon, the detailed verification procedures are developed. As the name states, the procedures should be very detailed, as they need to be repeatable as written and nothing should be inferred. The detailed verification procedure includes detailed equipment and environment setup procedures so the initial conditions of the verification method are also repeatable. In addition, detailed explanation of the pass/fail criteria and a method for capturing the final result are included.

Verification is performed throughout development. Early verification ensures system requirements trace to the stakeholder requirements, and therefore needs. Product verification checks that developed system items are meeting system requirements. Usually there is final system verification to ensure the whole system and its elements adhere to the system requirements. In other words, the systems engineer is constantly checking to make sure what is being done is what is supposed to be done.

5.7   VALIDATION

Next along the Vee model comes the concept of validation. As we stated in the previous section, the Vee model does not intend to dictate that verification must occur before validation (Figure 5.3). Actually, at the start of the system development, there are no requirements to use in verification. The first artifacts are the stakeholder requirements, which must be validated to ensure they represent the needs of the stakeholders.

As we said, validation answers the basic and usually most important question: Was the right system built? Validation activities confirm the completed system satisfies the stakeholders' needs. It is important for systems engineers to recognize that it is possible for all system requirements to pass verification, discussed in the previous section, yet the system does not pass validation. This situation is like a child getting socks for his birthday. He may need new socks, but he is not happy about the gift. The socks are verified but not validated. Requirements tracing and early validation of stakeholder requirements attempt to mitigate this scenario, but if not done properly, stakeholders can end up with new socks on their birthday.

Early validation begins when stakeholder requirements are being developed. Validating the stakeholder requirements can be accomplished through concept of operations, prototypes, or use cases that are discussed with the stakeholders. One important lesson for systems engineers is to understand that it is their job to extract the true desires and needs from the stakeholders. Continuing our analogy, the stakeholders may say they need something for their feet, but they really mean they want nice sneakers and not a pair of socks. The systems engineers must ensure they fully understand and are correctly interpreting the desires and needs of the stakeholders and not just writing down what the stakeholders say.

As an example of early validation, the systems engineer develops use cases and concept of operations based on interactions with the stakeholders. The systems engineer meets often with the stakeholders to validate that the true need and correct operational scenarios are captured. The stakeholders are asked whether the operational scenarios depicted in the use cases meet the expectations of the stakeholder. Let us say the stakeholders have expressed a need to get their dishes clean.

In this example, the systems engineer documents details of the use case "clean dishes," where dirty dishes are cleaned and sanitized. This use case is based on the stakeholder requirement "It needs to wash and dry kitchen items" from Section 2.1.1. The systems engineer sits down with the stakeholder and walks him or her through the use cases. During this process, the systems engineer may realize that the stakeholder not only wants clean dishes, but also wants them spotless. This unspoken requirement must be documented and added to the use cases. Once the stakeholders agree that the scenarios address their needs, the use cases are considered validated. Going forward, the systems engineer can refer back to these validated items to make sure the right system is being built.

The system is built and verified against the requirements. Validation plans are developed in advance for when the system is ready, usually after final verification is complete. This activity may be called end product validation, to differentiate it from early validation. The validation plans outline how the working system will be validated. These plans can incorporate use cases and concept of operations that were developed during early validation. The validation plans are reviewed and approved by the stakeholders. Detailed validation procedures are developed from the plan. The procedures should be detailed enough to be repeatable since validation of a system can occur with multiple stakeholders.

Using the dishwasher example above, a select group of customers are asked to install the dishwasher in their home and then asked to use the dishwasher for a set amount of time (e.g., a month or 6 weeks). The customer is asked to provide feedback on specific validation criteria, such has how clean the dishes were after usage, if there were any mechanical issues with the dishwasher during the validation period, what type of environment it was installed in, etc. Once the engineering team collects all the validation feedback, they can determine if the product passes validation or not.

A frequent mistake is that engineers believe they alone are able to perform final validation of a system. But, final validation can *only* be accomplished with the final stakeholders or customers. One last note of caution is that even though a system may meet every captured stakeholder requirement, the final system may not pass validation. This may be surprising, but it can happen for many reasons, including missing stakeholder requirements, misinterpreted stakeholder requirements, missing stakeholders (folks that were never asked for input but should have been), wrong stakeholder requirements, technology advancements, etc. In other words, end product validation may fail due to improper early validation. As we conclude this section on validation, it should occur to you that we have made a full circle and have come back to discussing the very first step in system development, stakeholder requirements. If you get the very first step wrong, that of developing stakeholder requirements, then you almost never get the last step right, which is the end goal of system validation.

6 Systems Engineering Management

Systems engineering is much more than design and engineering of a system. It encompasses the management associated with engineering a system also. Although some colleges offer a full degree in engineering management, an overview of a few select areas of management related to systems engineers are described as follows (please note this list may not be exhaustive).

While the project or program manager is responsible for the overall management of the project, the systems engineer has some management duties also. Project planning is the tactical and strategic means of defining system problems, forecasting conditions, and coordinating program elements to maximize program focus on providing superior products and services (Forsberg et al. 2005). It provides the guidance and tools to track and manage the technical aspects of program activity, monitors and controls the technical progress of a project, and tailors program-specific processes to optimally satisfy program needs. In contrast, project management directs the overall planning, organizing, securing, and managing of resources to bring about the successful completion of specific program goals and project objectives.

Project assessment and control ensures a project performs according to the project plans, schedules, and within the projected budget while meeting the technical objectives and expectations of the stakeholder. It is a risk-reduction approach and promotes continuous communication between the systems engineers, project managers, and any and all relevant stakeholders. Variances, potential risks, and any trends are types of information generated through this activity.

Decision analysis provides information for decision makers, such as the project managers, systems engineers, and management, to pick the best course of action for the design and development of a system. An element of decision analysis is the trade study activity. Trade studies identify the most balanced technical solutions among a set of proposed viable solutions. A trade study may be performed to choose the best set of requirements, architecture, design, and solution options, just to name a few. Anywhere there is a choice between two or more options, a trade study may be performed to gather the best information for making the decision.

Risk management is an activity within project management, but it is just as important to systems engineering. It identifies and analyzes the uncertainties of achieving program objectives related to a system and develops plans to reduce the likelihood or consequences of those uncertainties. While the project manager may focus more on schedule and cost, the systems engineer emphasizes the technical aspects of the project. However, it is incorrect to say that the systems engineer is not concerned at all with schedule and cost. Risk management provides the proper balance between issues, opportunity, and risks. A risk is an uncertain event or situation with a realistic probability of occurring that has an impact on the successful accomplishment of one or more project objectives. When the impact is beneficial, we call the risk an opportunity. Risks come in several varieties, such as cost risks, schedule risks, technical risks, and program risks. Although everyone on a project to engineer a system needs to be cognizant of risks, it is the technical risks that mainly fall under the purview of the systems engineer. When an event or situation that impacts the successful accomplishment of project objectives has occurred or is certain to occur, this event is called an issue instead of a risk.

Considered its own discipline in some settings, configuration management establishes and maintains consistency of a system's performance and functional and physical attributes with its concept, requirements, design, and architectural information throughout its life span. If there is not a separate configuration manager, the systems engineer is responsible for handling the configuration management aspects of the project.

Information management collects, manages, stores, and distributes all information, including raw data, pertaining to a particular project and system. It ensures the correct information is available when needed without compromising the information's security or value.

Requirements management has been introduced in Chapter 5, but it may be considered a management process more than a technical process. It tracks and controls all requirements and their associated artifacts. A requirement is an essential characteristic, condition, or capability that shall be met or exceeded by a system or component to satisfy the needs of the stakeholders. While the design of a system focuses on developing the requirements, requirements management ensures the captured requirements are managed to prevent loss and conflict. There is an overlap with configuration management, and in practice the two areas may be combined.

Interface management helps to ensure that all the pieces of the system work together to achieve the system's goals and continue to operate together as changes are made to the system. In practice, interface management is an integral part of requirements management, but the importance of considering interfaces causes some systems engineers to call it out as its own management area.

Technical performance measurement is a process to continuously assess and evaluate the adequacy of the system's architecture and design, as they evolve, to satisfy requirements and objectives. It quantitatively identifies potential design deficiencies and monitors progress relative to satisfying requirements. The result of this activity is an input to the risk management for the project.

Business process management is a holistic approach for improving processes, and may also be called quality management or continuous process improvement. In other words, business process management is an approach to improve the way systems engineering, as well as the other disciplines, is performed within an organization. These processes are a set of logically related tasks performed to achieve a defined organization outcome. It is an approach for redesigning the way work is done to better support an organization's mission and reduce costs by aligning all aspects of an organization with the wants and needs of stakeholders through analysis and design of workflows within and between organizations.

Many of the discussed management areas have formal processes developed and captured. These system life cycle processes can be found in one of the standards for systems engineering, *Systems Engineering: System Life Cycle Processes* (ISO-15288 2008).

7 Tools Used in Systems Engineering

In other disciplines, a tool may be a device that one can hold to accomplish a specific task. For systems engineering, tools are more conceptual and usually take the form of graphical devices or possibly even software applications. Some of the more common systems engineering tools will be discussed in brief.

The Integrated Definition for Function Modeling (IDEF0)—alternatively a compound acronym, ICAM Definition for Function Modeling, where ICAM is an acronym for integrated computer-aided manufacturing—is a process for modeling how inputs are transformed into outputs via some function. In other words, systems engineers use it primarily to explore requirements and interfaces. The resulting artifacts are called IDEF0 diagrams. An IDEF0 diagram can represent any level of system abstraction, and at least two diagrams are needed per system. The first IDEF0 diagram, known as page A-0, depicts the context diagram with the inputs, controls, outputs, and mechanisms for the top-level function of the system. This diagram establishes the scope and boundaries of the system and indicates interacting external systems. The remaining IDEF0 diagrams represent a decomposition of a function from a higher level of abstraction, starting with the function identified in A-0 (Buede 2000).

House of quality function deployment or quality function deployment (QFD) is a method to facilitate the transformation of stakeholder expectations into requirements. In many cases, these requirements are performance requirements, although they do not have to be. Often these stakeholder expectations are subjective and the QFD is a tool to translate them into well-defined terms (Blanchard and Fabrycky 1998). For example, the stakeholders may require a system to be easy to use or have an aesthetic interface. While most people would have a good idea of what is requested, they are very subjective requirements. The QFD is a way to map these subjective requirements to concrete, testable requirements. The concrete requirements can be validated with the stakeholder to ensure the interpretation of easy or aesthetic matches the stakeholders' expectations.

The Unified Modeling Language (UML) is a standardized, general-purpose modeling language with extensions for systems engineering, the Systems Modeling Language (SysML). SysML supports the specification, visualization, analysis, design, verification, and validation of a broad range of systems and systems of systems. Both UML and SysML consist of a series of integrated graphical tools. Although we will not review all the graphical tools, we will present a few that are common to both UML and SysML.

A context diagram shows the system boundaries, external entities that interact with the system, and the relevant information flows between these external entities and the system.

The use case diagram is a type of behavioral diagram defined by and created from a use case analysis. Its purpose is to present a graphical overview of the functionality provided by a system in terms of actors, their goals (represented as use cases), and any dependencies between those use cases. The main purpose of a use case diagram is to show what system functions are performed for which actor. Roles of the actors in the system can be depicted.

The sequence diagram is an interaction diagram that shows how processes operate with one another and in what order. It is a construct of a message sequence chart. Sequence diagrams are sometimes called event diagrams, event scenarios, and timing diagrams.

The activity diagrams are graphical representations of workflows of stepwise activities and actions with support for choice, iteration, and concurrency. They show the overall flow of control and can be used to describe the business and operational step-by-step workflows of components in a system.

Although used in both UML and SysML, the functional flow block diagrams have existed as systems engineering tools on their own. They provide a hierarchical decomposition of a system's functions. The diagrams illustrate the control structure that dictates the order in which the functions can be executed at each level of decomposition (Buede 2000).

In addition to graphical tools, there are some software tools available to assist the systems engineer. Requirement management tools are a set of software tools that have been developed to aid in managing requirements. As discussed earlier, there are many pieces of information and relationships (tracing) that must be managed for requirements, and these tools help facilitate that management. The requirement management tools are not a substitute for good systems engineering, just as computer-aided drafting programs are not a replacement for good mechanical engineering. These tools require a requirement development process so that the appropriate information is captured and updated in accordance with a documented process. Usually these tools allow for creation of requirements specification documents, requirements tracing information, associated requirement information, verification plans, verification tracing, and a multitude of reports.

8 Systems of Systems and Their Challenges

Up until now we have been discussing systems engineering as it applies to traditional systems. Remember from Chapter 1, a (traditional) system is a set of interrelated parts that work together to accomplish a common purpose. But, what happens when those parts are actually systems in their own right? We call the overall system a system of systems (SoS). Experience has shown that SoSs bring their own set of challenges that differentiate them from traditional systems, at least in several respects. A good systems engineer needs to identify SoSs so as to avoid applying traditional systems engineering to a situation where it may not work as expected.

At a minimum, a SoS is a type of system composed of other systems. An often cited depiction of a SoS describes a collection of component systems with two additional properties. Each component system must have its own purpose independent of the other systems, and the component systems must maintain their managerial independence (Maier 1998). However, many examples of SoSs, at least the ones that create the most challenges, possess some additional characteristics. Currently, there are arguments among system scientists over the correct set of characteristics, and some argue that all systems exhibit variations of these characteristics to some degree. Suffice it to say that a SoS is fundamentally a system, as its name suggests, that exists as a composite of independent systems. Yet additionally, they maintain connections and collectively contribute to the goal of the overall SoS. The complexity of these interacting, constituent systems causes the emergence of behavior that cannot be traced to any of the constituent systems (Baldwin and Sauser 2009; Baldwin et al. 2011). Table 8.1 presents an overview comparison of traditional systems to SoSs based on one set of SoS characteristics first identified by Boardman and Sauser (2006).

So why worry about a SoS any differently than any other system? Whether the result of scope or unique characteristics, a SoS has additional challenges. These difficulties include the following:

- Autonomy, or independence, of the many constituent systems results in management issues.
- The SoS may not have its own set of requirements, although it may have its own set of goals. If requirements do exist, they are most likely very ambiguous.
- Interaction of systems grows exponentially as constituent systems are added to the SoS.
- Increased interaction causes interfaces to conflict and reduces the quality of the interface documentation.
- Confusion over scope of constituent systems as well as overall SoS causes management, as well as risk management, issues.

TABLE 8.1

Differences between a Traditional System and a System of Systems

Traditional System	System of Systems
Overall system is autonomous, but its parts are not	Overall SoS is autonomous, and its constituent systems work as well independently as part of the SoS
Parts of a system collaborate only to the extent they are designed	Constituent systems collaborate as needed to help each other reach the goals of the SoS
Parts of a system are statically connected	Constituent systems may be dynamically connected, joining and separating from the SoS as needed
Each system is unique but its parts may be mass-produced	SoS is composed of a diversity of constituent systems
A system is more than the sum of its parts, but its functionality is understood	A SoS exhibits emergent functionality that cannot be traced to any particular constituent system

- Management problems are caused by diverse configurations (INCOSE 2010).
- SoSs evolve over time, and therefore engineering is never finished (Maier 1998).
- Interoperability of constituent systems causes changes in one system to have unexpected impacts on other systems (ODUSD(A&T)SSE 2008).
- Functionality emerges from the connections between constituent systems (DeLaurentis and Callaway 2004).
- Test and validation are extremely complicated (ODUSD(A&T)SSE 2008).

This list of challenges may not be inclusive, as the emergent nature of a SoS can cause any number of challenges. For example, one obvious challenge when dealing with any constituent system in a SoS is risk management. Due to the interconnected nature of the systems, a change to one system may ripple through other systems. However, risk management traditionally focuses on the system of interest and generally lacks authority to mitigate risks outside of its domain.

Another important challenge is the difference in integration of systems into a SoS from traditional integration of parts into a system. It is difficult enough to integrate the elements of a system, but the problem is compounded when those parts are actually autonomous systems. Unfortunately, there is no clear-cut answer to this problem, but it may be beneficial for the systems engineer to know what difficulties lie ahead.

9 Learning More about Systems Engineering

As you have found, systems engineering is a way of thinking about a system of interest in a holistic manner.

Although there are many facets of systems engineering, it basically can be explained as focusing on addressing *why* a system is needed, *what* the system must do, and then *how* the system will accomplish the tasks over the entire life of the system—from conception to disposal. There are many times when the systems engineer must continually question why something is needed and what the system will do to address this need. Why does the system need a certain trait? What will the trait do to address the system's reason for being? And only after these basic questions are answered, how will the system implement these traits?

So, systems engineers are skilled in identifying the true problem or need before attempting to determine any solution, since the answer to the wrong problem is not really a solution. While this sounds intuitive, it takes a lot of practice. In industry, systems engineers have acquired a lot of knowledge about how systems, and their parts, interact.

However, unlike the classical engineering disciplines, systems engineering is primarily comprised of best practices rather than mathematical models and theory. As presented in this booklet, systems engineering requires the eliciting of stakeholder requirements and understanding how those stakeholders want to incorporate a new system into their existing environment. For example, how would a person change the way he or she listened to music when the MP3 player was introduced? If the new system is part of a broader enterprise, politics can be involved. For instance, there may be politics involved with what contractor is selected to design the new system. No amount of mathematics or theory can begin to deterministically describe that issue. But, there are aspects that do behave in a deterministic manner, such as characteristics of a network system or the behaviors of a system of systems. While network theory is pretty well established, system of systems theory is still in its infancy as far as mathematical modeling. Wherever possible, systems scientists need to strive toward modeling and developing solid theory.

If this booklet has caused you to become more interested in systems engineering, we suggest you explore the subject more. Begin thinking about a system as a black box, and consider what external interfaces interact with this black box. Then, think about a part of the same system and consider it a black box. Ask, what does this part of the system connect to? Maybe you want to explore drawing your system of interest using a systems engineering notation such as Integrated Definition for Function Modeling (IDEF0) or Systems Modeling Language (SysML). While there are tools available to help do that, simply using paper and pencil, or PowerPoint/Visio, will help you better reason about the problem at hand. Maybe you can take a class or just read more on the systems engineering approach. There are a number of great books included in the references at the end of this booklet. Finally, consider joining a systems engineering professional society. We suggest the International Council on Systems Engineering (INCOSE), but there are other societies with a systems focus, such as the Institute of Electrical and Electronics Engineers (IEEE) and the American Society for Engineering Management (ASEM). However you choose to proceed, we hope you evaluate your options from a holistic viewpoint, taking a systems approach.

References

Ackoff, R. L. 1981. Creating the Corporate Future: Plan or Be Planned for. New York: John Wiley & Sons.

Alexander, Ian, and Richard Stevens. 2002. *Writing Better Requirements*. London: Addison Wesley.

Bahill, A. Terry, and Bruce Gissing. 1998. Re-Evaluating Systems Engineering Concepts Using Systems Thinking. *IEEE Transactions on Systems, Man, and Cybernetics: Part C: Applications and Reviews* 28(4): 516–527.

Baldwin, W. Clifton, Wilson N. Felder, and Brian J. Sauser. 2011. Taxonomy of Increasingly Complex Systems. *International Journal of Industrial and Systems Engineering* 9(3): 298–316.

Baldwin, W. Clifton, and Brian Sauser. 2009. Modeling the Characteristics of System of Systems. In *2009 IEEE International Conference on System of Systems Engineering (SoSE)*. Albuquerque, NM: IEEE, 1–6.

Blanchard, Benjamin S., and Wolter J. Fabrycky. 1998. *Systems Engineering and Analysis*. 3rd ed. Upper Saddle River, NJ: Prentice-Hall.

Boardman, John, and Brian Sauser. 2006. System of Systems—The Meaning of Of. In *Proceedings of the 2006 IEEE/SMC International Conference on System of Systems Engineering*, 118–123. Los Angeles: IEEE. doi:10.1109/SYSOSE.2006.1652284.

Bone, Mary. 2008. Cyclone Process: Dealing with Vague Requirements. Presented at INCOSE Conference on Systems Engineering Research, Los Angeles.

Booch, Grady, Robert Maksimchuk, Michael Engle, Bobbie Young, Jim Conallen, and Kelli Houston. 2007. *Object-Oriented Analysis and Design with Applications*. 3rd ed. Boston: Pearson Education.

Buede, Dennis M. 2000. *The Engineering Design of Systems: Models and Methods*. New York: John Wiley & Sons.

Charette, Robert. 2005. Why Software Fails. *IEEE Spectrum*.

Christel, Michael, and Kyo Kang. 1992. *Issues in Requirements Elicitation (CMU/SEI-92-TR-012)*. Pittsburgh, PA: Software Engineering Institute, Carnegie Mellon University.

DAU. 2001. *Systems Engineering Fundamentals*. Fort Belvoir, VA: Defense Acquisition University Press.

DeLaurentis, Daniel A., and Robert K. Callaway. 2004. System-of-Systems Perspective for Public Policy Decisions. *Review of Policy Research* 21(6): 829–837.

EIA-632. 1999. *Processes for Engineering a System*. Standard. Arlington, VA: Government Electronics and Information Technology Association.

Forsberg, Kevin, Hal Mooz, and Howard Cotterman. 2005. *Visualizing Project Management: Models and Frameworks for Mastering Complex Systems*. 3rd ed. Hoboken, NJ: John Wiley & Sons.

INCOSE. 2006. INCOSE—A Consensus of the INCOSE Fellows. http://www.incose.org/practice/fellowsconsensus.aspx.

INCOSE. 2010. *INCOSE Systems Engineering Handbook: A Guide for System Life Cycle Processes and Activities*, ed. Cecilia Haskins. Version 3.2, INCOSE-TP-2003-002-03.2. Seattle, WA: International Council on Systems Engineering.

ISO-15288. 2008. *Systems Engineering: System Life Cycle Processes*. Standard. Geneva, Switzerland: International Organization for Standardization.

Kelly, M. J. 1950. The Bell Telephone Laboratories: An Example of an Institute of Creative Technology. *Proceedings of the Royal Society B*. 203(287–301).

Maier, Mark W. 1998. Architecting Principles for System-of-Systems. *Systems Engineering* 1(4): 267–284.

McManus, J., and T. Wood-Harper. (2007). Understanding the Sources of Information Systems Project Failure. *Management Services* 51(3): 38–43.

Mil-Std-499, Military Standard: System Engineering Management (17 Jul 1969).

Mil-Std-961E, Military Standard: Defense and Program-Unique Specifications Format and Content (1 Aug 2003).

NASA. 2007. *NASA Systems Engineering Handbook*. Washington, DC: National Aeronautics and Space Administration.

ODUSD(A&T)SSE. 2008. *Systems Engineering Guide for Systems of Systems*. Version 1.0. Washington, DC: Office of the Deputy Undersecretary of Defense for Acquisition and Technology. http://www.acq.osd.mil/sse/docs/SE-Guide-for-SoS.pdf.

Shaw, Mary. 1984. Abstraction Techniques in Modem Programming Languages. *IEEE Software* I(4): 10–26.

Standish Group. 1994. *Chaos*.

Zhang, Qian, and Armin Eberlein. 2002. Deploying Good Practices in Different Requirements Process Models. Presented at Proceedings of the 6th IASTED International Conference on Software Engineering and Applications, Cambridge, MA.

Index

Printed in the United States
by Baker & Taylor Publisher Services